HRW

Chapters 1-3

ADVANCED
ALGEBRA

TEACHING RESOURCES

Y0-CDJ-573

explore

communicate

APPLY

$g(\theta)=15\sin(200\theta)$

$h(t)=-16t^2+27t+240$

HOLT, RINEHART AND WINSTON
Harcourt Brace & Company
Austin • New York • Orlando • Atlanta • San Francisco • Boston • Dallas • Toronto • London

TO THE TEACHER

HRW Advanced Algebra Teaching Resources contains blackline masters that complement regular classroom use of *HRW Advanced Algebra*. They are especially helpful in accommodating students of varying interests, learning styles, and ability levels. The blackline masters are conveniently packaged in four separate booklets organized by chapter content. Each master is referenced to the related lesson and is cross-referenced in the *Teacher's Edition*.

- **Practice Masters** (one per lesson) provide additional practice of the skills and concepts taught in each lesson.
- **Enrichment Masters** (one per lesson) provide stimulating problems, projects, games, and puzzles that extend and/or enrich the lesson material.
- **Technology Masters** (one per lesson) provide computer and calculator activities that offer additional practice and/or alternative technology to that provided in *HRW Advanced Algebra*.
- **Lesson Activity Masters** (one per lesson) connect mathematics to other disciplines, provide family involvement, and address "hot topics" in mathematics education.
- **Chapter Assessment** (one multiple-choice test per chapter and one free response test per chapter)
- **Mid-Chapter Assessment** (one per chapter)
- **Assessing Prior Knowledge and Quiz** (One Assessing Prior Knowledge per lesson and one Quiz per lesson)
- **Alternative Assessment** (two per chapter) is available in two forms, one which entails concepts found in the first half of the chapter and the other which entails concepts found in the second half of the chapter.

Copyright © by Holt, Rinehart and Winston, Inc.

All rights reserved. No part of this publication may be reproduced or transmitted in any form or by any means, electronic or mechanical, including photocopy, recording, or any information storage and retrieval system.

Permission is granted for the printing of complete pages for instructional use and not for resale by any teacher using HRW ADVANCED ALGEBRA: EXPLORE, COMMUNICATE, APPLY.

Developmental assistance by B&B Communications West, Inc.

Printed in the United States of America

ISBN 0-03-095390-1

2   3   4   5   6   7   066   99   98   97

TABLE OF CONTENTS

Practice & Apply
1.1 Tables and Graphs of Linear Equations

Are the variables in each table linearly related?

1.

x	y
-2	-3
-1	-1
0	1
1	3
2	5

2.

x	y
1	-3
3	5
5	21
7	45
9	77

3.

x	y
0	3
-2	6
-4	9
-6	12
-8	15

4.

x	y
1	1
4	2
9	3
16	4
25	5

_____   _____   _____   _____

Which of the following equations are linear?

5. $\frac{3}{4}y + \frac{1}{2}x = 1$

6. $xy = 2$

7. $y = 3x^2 - 2$

8. $4x = \frac{2}{5}y + 3$

_____   _____   _____   _____

Use the graph to answer Exercises 9-13.

9. As x increases by 1, what is the difference in consecutive y-values?

_____

10. What y-value would be associated with an x-value of 0?

_____

11. Write an equation for the relationship.

_____

12. Describe why the equation is linear.

_____

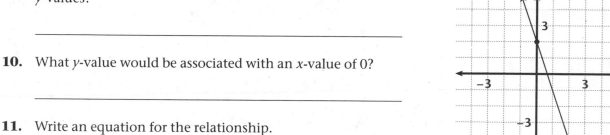

In Exercises 13–16, match each table of values with its equation.

A.

x	-2	-1	0	1	2
y	-6	-3	0	3	6

B.

x	-6	-3	0	3	6
y	-2	-1	0	1	2

C.

x	-12	-6	0	6	12
y	4	2	0	-2	-4

D.

x	-2	-1	0	2	4
y	-9	-6	-3	3	9

13. $y = 3x - 3$ _____

14. $y = -\frac{1}{3}x$ _____

15. $y = 3x$ _____

16. $y = \frac{1}{3}x$ _____

Practice & Apply
1.2 Exploring Slopes and Intercepts

Name the slope *m* and the *y*-intercept *b* of each line.

1. $y = 0.2x - 3$ _____

2. $y = -4x + 7$ _____

3. $y = -x$ _____

4. $y = \frac{1}{2}x - 6$ _____

5. $y = 5$ _____

6. $y = -\frac{3}{4}x + 8$ _____

Write a linear equation with the indicated slope *m* and *y*-intercept *b*.

7. $m = 3; b = 4$ _____

8. $m = -2; b = 0$ _____

9. $m = -\frac{1}{4}; b = 5$ _____

10. $m = 0; b = -6$ _____

For Exercises 11–14, (a) find the slope of a line passing through the indicated points, and (b) write the equation of the line in slope-intercept form.

11. $(2, 3)$ and $(5, 9)$ _____

12. $(-1, 4)$ and $(3, -8)$ _____

13. $\left(\frac{1}{3}, -2\right)$ and $\left(2, -\frac{1}{3}\right)$ _____

14. $(-3, -5)$ and $(-1, -2)$ _____

15. Which equation has a 0 slope? _____

A. $y = 4x - 4$ **B.** $x = 4$ **C.** $y = 4$ **D.** $y = -4x + 4$

16. Which equation does not have the same slope as $3x + y = 6$? _____

A. $-3x - 6 = -y$ **B.** $6x + 2y = 12$ **C.** $x + \frac{1}{3}y = 6$ **D.** $-6 + y = -3x$

17. To find the slope of the line through $(-1, 9)$ and $(2, 3)$, Jane found the slope as $\frac{9 - 3}{-1 - 2}$. Paul found the slope as $\frac{3 - 9}{2 - (-1)}$. Compare the two answers.

_____

The linear equation $y = 9.97 + 1.05x$ models the extent of air and water pollution where x is the amount of industrialization of a geographical region and y is the amount of air and water pollution.

18. Use your own graph paper to graph the equation.

19. Suppose the geographical region has an industrialization value of 80. What would the pollution value be? _____

Practice & Apply
1.3 Scatter Plots and Correlation

The manager of a department store conducted a survey among customers holding charge cards. He surveyed 12 customers to estimate the amount he or she charged at the store during the past month. The data from his survey is shown in the table.

Actual	75	85	39	87	18	74	31	58	9	64	75	80
Estimate	75	99	40	90	24	68	40	64	10	40	82	75

1. Enter the data, and display it in a scatter plot.

2. What correlation best describes this data: positive, negative, or zero? _____

3. Are there any outliers? If so, name them. _____

4. Find the correlation coefficient r to the nearest tenth. _____

A teacher is doing a study of the relationship between student grades and the hours worked on their part-time job. The teacher collects data from six students. The number of hours worked per week and the grade-point averages of these students are organized in the table.

Hours/week x	20	22	10	30	7	25
GPA y	2.5	3.0	3.5	2.7	3.2	1.5

5. Enter the data, and display it in a scatter plot.

6. What is the general shape of the scatter plot? _____

7. Find the correlation coefficient r to the nearest tenth. _____

8. Are there any outliers? If so, name them.

   _____

9. What is the equation of the line of best fit? _____

10. Do you think the scatter plot shows that part-time work is the cause of low grades? Explain.

   _____

Practice & Apply
1.4 Direct Variation and Proportion

For each of the following values of x and y, y varies directly as x. Find the constant of variation and write the equation of direct variation.

1. y is 12 when x is -8

2. y is -9 when x is 15

3. y is 2.6 when x is 1.3

_____   _____   _____

Write an equation of direct variation that relates the two variables.

4. The circumference C of a circle varies directly as the length of the diameter d.

_____

5. The weekly income I varies directly as the number of hours h worked.

_____

For Exercises 6–10, refer to the graph.

6. Explain why the graph represents a direct variation.

_____

_____

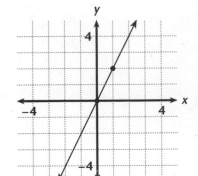

7. What is the constant of variation? _____

8. Write the equation of direct variation.

_____

9. If $x = 3$ when $y = 6$, what is y when $x = -2$? _____

10. Compare the ratio of the y-value to the x-value in at least 3 ordered pairs. What did you notice?

_____

The tables represent ordered pairs of an equation. Give the constant of variation and write the equation of direct variation.

11.

x	-10	-5	5	10	20
y	4	2	-2	-4	-8

12.

x	-2	-1	0	1	2
y	2	1	0	-1	-2

_____   _____

NAME _____ CLASS _____ DATE _____

Practice & Apply
1.5 Solving Equations

Solve the following equations by graphing.

1. $3x = 4x - 1$ _____
2. $2x - 7 = 11$ _____
3. $8 - 5x = -7$ _____

4. $9x + 1 = 19$ _____
5. $-3x - 6 = 12$ _____
6. $-4x + 2 = 10$ _____

7. $\frac{2}{3}x = 9$ _____
8. $\frac{4}{5}x + 3 = 7$ _____
9. $-\frac{1}{2}x + 3 = 5$ _____

10. $4x - 1 = 2x - 3$ _____
11. $\frac{2}{3}x + 2 = \frac{1}{3}x - 1$ _____
12. $2x + \frac{1}{4} = \frac{3}{2}x + \frac{1}{2}$ _____

Use your graphics calculator to estimate, to the nearest tenth, the solution to each of the following equations.

13. $\frac{3}{4}x + 5 = 7$ _____
14. $5x = \frac{2}{3}x + 2$ _____
15. $8\left(\frac{1}{4}x - \frac{3}{4}x\right) = 2x + 7$ _____

16. The sum of the measures of two complementary angles is 90°. The measure of one complementary angle is 10° more than three times the measure of its complement. Find the measure of each angle. _____

17. Given that $y = 2x - 3$, use substitution to solve $4x - y = 12$ for x. _____

18. Given that $x = 2y + 1$, use substitution to solve $3x - 2y = -3$ for y. _____

19. Solve $A = P + Prt$ for P. _____
20. Solve $C = \frac{5}{9}(F - 32)$ for F. _____

21. Solve $a_n = a_1 + (n - 1)d$ for d. _____
22. Solve $A = \frac{1}{2}h(b_1 + b_2)$ for h. _____

Solve each equation for x.

23. $3x - 4(3x - 8) = 35$ _____
24. $6(3x - 4) = 5(4x - 2)$ _____

25. $2(2 - 3x) + 20 = 4x - (5x - 4)$ _____
26. $4\left(\frac{1}{3}x + 2\right) = 2(x - 6) + 8$ _____

27. An electronics store marks up each TV set it sells 60% above wholesale price. What is the wholesale price of a TV set that retails for $240? _____

28. Tickets for a play cost $12 for orchestra seats and $9 for balcony seats. If the total receipts from 400 tickets were $4350, how many of each type of ticket were sold?

_____

NAME _____ CLASS _____ DATE _____

Practice & Apply
1.6 Solving Inequalities

Write an inequality that describes the graph.

1.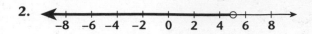

_____

2.

_____

Solve and graph each inequality.

3. $2x - 1 > 5$

_____

4. $x + 12 \leq 3x + 2$

_____

5. $\frac{7x - 1}{3} \geq x - 6$

_____

6. $-16 + 6x > 10 - 7x$

_____

7. $9 - (x - 1) < -2x + 10$

_____

8. $7 - 4x \leq 4x - 9$

_____

9. $y < -3x + 2$

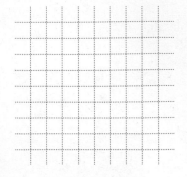

_____

10. $y - 1 > \frac{3}{2}x$

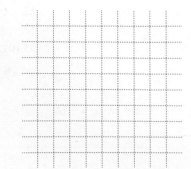

_____

11. $\frac{x - 3}{4} \geq \frac{y}{2}$

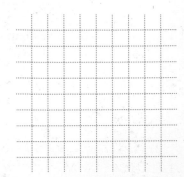

_____

HRW Advanced Algebra

Enrichment
1.1 Better Choice

Complete the table to answer the question. Then write an equation.

1. Underwater Repair Plumbing Services charges $60 for a house call plus $56 per hour labor. Waterworks Plumbers charges $75 for a house call plus $52 per hour labor. Each has estimated your job to last 10 hours. Which service should you choose based on cost?

Hours	Underwater	Waterworks
0	60	75
1	116	127
2	172	179

_____

_____

2. East High School wants to hire West Wind Band for a dance. The band charges $300 plus $2 per person in attendance. The school will charge $5 per ticket and wants to know the minimum number of students needed in order to break even.

Number of students	Profit
0	−300
10	−270
20	−240

_____

_____

3. To mail a package to Zone 3 costs $1.30 for the first 1.5 lb of bound printed matter, then an additional $0.04 for each 0.5 lb, up to a total of 5 lb. What is the cost of mailing 5 lb of bound printed matter?

Weight (lb)	Cost
1.5	$1.30
2.0	$1.34

_____

_____

4. An owner of a mobile home park charges a $100 fee for moving a trailer onto a lot plus $80 per month rent. What is the cost of living at the park for the first year?

Month	Cost
0	100
1	180

_____

_____

Enrichment

1.2 Moving Around the Maze

For each square in the maze, find the slope of the line that contains the two
given points. To find your way out of the maze, enter the maze where indicated,
then move one square vertically or horizontally. The square you move to must
have a slope greater than that of the square you are currently in.

START					
(1, 10) (2, 2)	(2, 15) (4, 5)	(−1, 68) (3, 20)	(3, 2) (−1, 20)	(6, −14) (−4, 16)	(1, 2) (2, 10)
(3, 4) (2, 14)	(−5, 2) (−2, −10)	(4, 3) (−1, 18)	(3, 4) (−2, 4)	(−8, 12) (−4, −6)	(4, 17) (−6, −18)
(−10, 17) (−8, −4)	(−8, 8) (−9, 17)	(5, 6) (8, −10)	(9, −12) (10, −11)	(7, 6) (10, 8)	(8, 1) (−2, −4)
(−3, 10) (−7, 1)	(4, 10) (6, 14)	(9, 40) (−7, 10)	(9, 10) (7, 7)	(10, −9) (6, −10)	(15, 12) (7, 3)
(12, 2) (16, 12)	(6, 18) (−7, 5)	(12, 15) (3, 7)	(−1, 15) (−8, 9)	(−12, 2) (−4, 11)	(3, −1) (14, 5)
(7, 1) (8, 4)	(12, −1) (16, 13)	(7, −4) (11, 12)	(8, −10) (4, −27)	(7, 12) (5, 3)	(15, 11) (3, 4)
(−1, 10) (−4, 3)	(5, 14) (1, 1)	(2, 9) (10, 40)	(5, 16) (8, 10)	(4, 1) (12, 40)	(10, 10) (5, 15)
(1, −4) (2, −9)	(10, 12) (3, 7)	(18, 3) (7, −1)	(17, 8) (9, 2)	(5, 10) (4, 5)	(31, 2) (34, 18)
(15, 21) (22, 32)	(13, 9) (8, 13)	(5, 7) (12, 13)	(19, 10) (10, 5)	(21, 17) (19, 9)	(4, 2) (2, −10)

EXIT

Enrichment
1.3 Related Data

**Use your own graph paper to draw a scatter plot to represent
each set of data. Determine whether the correlation is positive,
negative, or none if it does not seem to exist and circle the
appropriate letter.**

1. Commercial Carrier Air Accidents   positive: P   negative: M   none: F

No. of Accidents	3	5	4	0	4	4	4	1	4	2	4	3	11	6	4	4
No. of Fatalities	78	160	351	0	4	233	15	4	197	5	231	238	278	39	62	33

2. U.S. Car Sales (in millions)   positive: I   negative: O   none: E

Consumer	6.1	5.6	5.3	6.1	6.6	7.1	7.7	6.7	6.8	6.4	5.8	4.5	4.6
Government	0.13	0.12	0.10	0.12	0.14	0.13	0.13	0.14	0.14	0.14	0.15	0.10	0.10

3. Car Sizes Sold (Percent of Total)   positive: O   negative: T   none: N

Small	38.8	39.1	37.9	37.6	38.4	37.6	38.6	35.2	35.7	32.9
Midsize	40.6	39.6	42.1	42.5	42.3	42.5	41.9	42.8	42.6	44.5

4. Production of Pig Irons and Raw Steel (million tons)   positive: L   negative: N   none: P

Pig Iron	46	53	65	77	66	88	91	101	68	59	44	48	56	56	55	49	52
Raw Steel	67	80	97	117	99	131	132	117	112	88	82	89	100	98	99	88	93

5. Zoo Attendance (millions)   positive: M   negative: S   none: I

Attendance	0.6	0.9	2.1	0.5	2.0	1.3	0.9	0.5	1.3	1.0	1.5	4.0	1.8
Species	200	500	670	206	300	761	563	329	310	413	596	339	500

6. College Football Win-Loss Record   positive: S   negative: N   none: A

Win	34	25	24	21	21	19	17	12	8
Loss	4	10	8	9	10	11	13	16	19

7. Vehicle Production Worldwide, 1992 (thousands)   positive: S: negative: R   none: E

Passenger Cars	220	280	23	814	180	930	12	154
Commercial Vehicles	41	5	4	257	900	600	6	170

8. What phrase is spelled out by your answers to Exercises 1–7? _____

 Enrichment
1.4 Direct Variation

Determine whether each table represents a direct variation. If it does not, shade in that region. What letter of the alphabet do the shaded areas form? _____

x	y
0	1
1	3
2	5

x	y
0	0
1	2
2	4

x	y
0	0
2	1
4	2

x	y
2	−4
4	−8
6	−12

x	y
4	1
8	3
12	5

x	y
0	4
1	7
2	10

x	y
0	0
1	3
2	6

x	y
0	0
4	1.2
8	2.4

x	y
2	0.4
4	0.8
6	1.2

x	y
2	1.4
4	1.8
6	2.2

x	y
0	0
1	0.4
2	0.8

x	y
2	−0.8
4	0.4
6	1.6

x	y
0	0
4	28
8	56

x	y
0	4
2	2
4	0

x	y
4	0.4
6	0.6
10	1.0

x	y
5	−1.5
10	−3.0
15	−4.5

x	y
4	−0.4
12	2.8
16	4.4

x	y
2	1.4
3	2.1
4	2.8

x	y
5	4.5
6	5.2
7	5.9

x	y
4	3.2
8	6.4
12	9.6

x	y
2	5
4	10
8	20

x	y
3	8.4
6	16.8
9	25.2

x	y
6	12.2
12	28.4
18	44.6

x	y
6	20.4
10	34
12	40.8

x	y
−1	−12
−2	−24
−3	−36

x	y
−1	−8
0	0
1	8

x	y
−1	4
0	0
1	−4

x	y
−1	4
0	1
1	−2

x	y
1	6
2	12
7	42

x	y
8	44.8
9	50.4
15	84.0

Enrichment
1.5 Graphing Equations

Match each equation with the graph that shows its solution.
Write the letter of the graph in the space provided.

1. $x = 2x - 4$

2. $2x = x + 6$

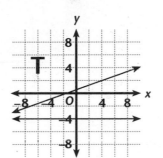

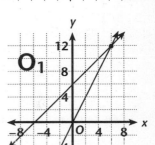

3. $\frac{1}{2}x + \frac{3}{2} = 2$

4. $\frac{1}{4}x - \frac{3}{4} = 1$

5. $-4 = \frac{1}{3}x + \frac{2}{3}$

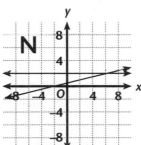

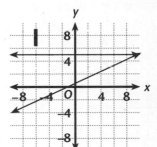

6. $5 = \frac{2}{3}x + \frac{1}{5}$

7. $-\frac{1}{4}x + \frac{3}{4} = 2$

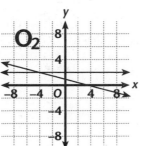

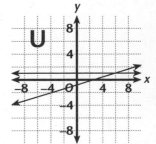

8. $\frac{1}{5}x + \frac{4}{5} = 2$

____ ____ ____ ____ ____ ____ ____ ____
1   2   3   4   5   6   7   8

Enrichment
1.6 Graphing Inequalities

Write an inequality for each situation. Then graph the inequality.

1. The number of males and females that Brandon can invite to his party can be no more than 15.

_____

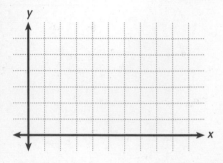

2. Cindy has 120 feet of fence. She wants to build a rectangular pen and is trying to decide on the pen's perimeter.

_____

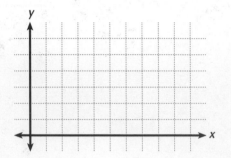

3. Mindy found some sweaters and skirts all on sale for $12 each. She has $100 to spend.

_____

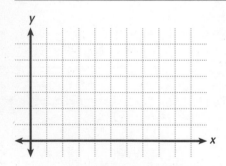

4. Calvin wants to plant some peas and green beans in his garden. He has room for 10 rows.

_____

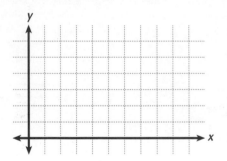

5. Jim and Julie are ordering centerpieces for a banquet. Centerpieces with daisies cost $24 each, and those with roses cost $30 each. They can spend up to $400.

_____

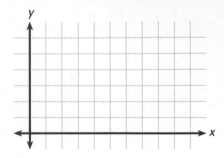

6. In order to go on a weekend band outing, band members must raise at least $1500. Candles sell for $3 each and popcorn buckets sell for $5 each.

_____

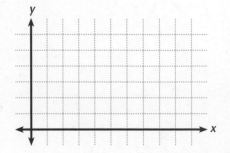

Technology
1.1 Tables, Equations, and Linearity

Suppose you are given $y = 0.5x^2 - 2$ and the question, "Is the relationship between x and y linear?" To answer the question, you can use a spreadsheet. In the spreadsheet shown, column C displays the differences of consecutive x-values and column D displays the differences of consecutive y-values.

	A	B	C	D
1	X	Y	X-DIFFERENCE	Y-DIFFERENCE
2	−2	0.0		
3	−1	−1.5	1	−1.5
4	0	−2.0	1	−0.5
5	1	−1.5	1	0.5
6	2	0.0	1	1.5

B2 contains .5*A2^2–2.
C3 contains A3–A2.
D3 contains B3–B2.

Use the FILL DOWN command to extend the formulas.

From the table, the consecutive differences of x-values are equal but the consecutive differences of y-values are not.

So, $y = 0.5x^2 - 2$ represents a nonlinear relationship between x and y.

Use a spreadsheet to determine if the given equation represents a linear or nonlinear relationship between x and y.

1. $3x - 5 = y$

_____

2. $y = -x^2 + 1$

_____

3. $y = 3$

_____

4. $2x - y = 5$

_____

5. $\frac{x}{5} - 6 = y$

_____

6. $x^2 - x = y$

_____

7. $y = \frac{1}{x}$

_____

8. $y = (2x + 1)(2x - 1)$

_____

9. $y = x^3$

_____

10. $2x + 7y = -14$

_____

Technology
1.2 Spreadsheets and Slopes

Some very simple formulas can generate some very interesting results.
Suppose you begin with the following rule.

$$\text{initial input} = 1 \qquad \text{output} = 2.5(\text{input}) + 3.5$$

Here is how to use a spreadsheet to generate subsequent input and output
values. Each value in column A from row 3 onward is obtained by taking
the value in column B on the previous row. For example, A3 = B2, A4 = B3,
and so on. Cell B2=2.5*A2+3.5, B3=2.5*A3+3.5, and so on.

	A	B
1	INPUT	OUTPUT
2	1.0000	6.0000
3	6.0000	18.5000
4	18.5000	49.7500
5	49.7500	127.8750
6	127.8750	323.1875

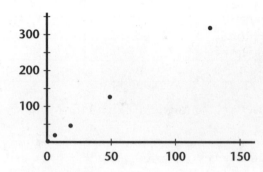

The graph generated by the spreadsheet program suggests that the points
generated by the table lie along a straight line.

**Use the indicated rows of the table to find the slope of the line
containing the points generated by the table.**

1. Rows 2 and 3

2. Rows 5 and 6

_____   _____

**Generate a table of values and graph for each rule. Find the
slope of the line containing the points from the table.**

3. initial input = 0
output = input −5

4. initial input = 4.7
output = 2(input) + 3.3

5. initial input = − 2.5
output = −7(input) −1.1

_____   _____   _____

6. initial output = −2
output = −2(input)

7. initial input = 4.1
output = 2(input) + 3.3

8. initial input = − 2.5
output = −7(input) + 1.1

_____   _____   _____

9. Find the slope of the line produced by the rule: input = 1 and
output = a(input) + 5.

_____

NAME _____ CLASS _____ DATE _____

Technology
1.3 Graphics Calculators and Line of Best Fit

The table shows data gathered in a survey about time spent studying and test scores.

Hours (h)	1	1	2	2	2	3	3	3	3	4
Score (s)	60	55	68	70	100	65	70	74	68	70

Hours (h)	4	4	5	5	6	6	6	6	7
Score (s)	80	88	92	75	85	94	100	98	95

The graphics calculator display shows the data for the line of best fit. Rounding the values of a and b, gives the equation of the line of best fit $s = 5.716h + 57.356$. A spreadsheet can be used to compare the actual data values to the values predicted by the equation. Hours are entered into column A and the score in column B. The line of best fit equation is entered into column C. Column D contains the percent change from the data values:

$$100 \times \frac{\text{equation value} - \text{data value}}{\text{data value}}$$

```
LinReg
 y=ax + b
 a=5.715652174
 b=57.35565217
 r =.734204457
```

	A	B	C
1	1	60	5.120
2	1	55	14.676
3	2	68	1.159
4	2	70	−1.731

Solve each problem or answer each question.

1. Which data values differ from the predicted values by more than 15%? _____

2. Which data values differ from the predicted by less than 5%? _____

3. Which predicted values agree with the actual data values? _____

4. Does the spreadsheet suggest any outliers? _____

5. Compare the average score for those who studied six hours with the predicted score.

   _____

6. Compare the average score for those who studied three hours with the predicted score.

   _____

7. Find the difference between each actual and each predicted score. _____

8. Find the average of the differences between actual and predicted values. _____

Technology
1.4 Spreadsheets, Direct Variation, and Proportion

When you are given a table of values for two variables, you can use a spreadsheet to see if they form a direct variation relationship. Set the spreadsheet up in such a way that columns A and B contain the table of values and column C contains the quotient $\frac{y\text{-value}}{x\text{-value}}$.

The spreadsheet for one table of values is shown. (Note that no quotient is given for the pair (0, 0)).

Since each entry in column C, except the first, is equal to each other entry in column C, you can conclude that the table of x-y pairs forms a direct variation relationship. Furthermore, 3.44 is the constant of variation.

	A	B	C
	X	Y	Y/X
1	X	Y	Y/X
2	0.0	0.00	
3	0.5	1.72	3.44
4	1.0	3.44	3.44
5	1.5	5.16	3.44
6	2.0	6.88	3.44
7	2.5	8.60	3.44
8	3.0	10.32	3.44
9	3.5	12.04	3.44

Extend each spreadsheet to calculate the quotient $\frac{y\text{-value}}{x\text{-value}}$. Is there a direct variation relationship between x and y? Write yes or no. If yes, give the constant of variation.

1.

	A	B
	X	Y
1	4.0	2.0
2	4.4	2.2
3	4.8	2.4
4	5.2	2.6
5	5.6	2.8

2.

	A	B
	X	Y
1	−3.5	−2.45
2	5.3	4.24
3	6.1	4.27
4	2.3	1.84
5	4.9	3.92

3.

	A	B
	X	Y
1	10.2	16.32
2	9.9	15.84
3	9.6	15.36
4	9.3	14.88
5	9.0	14.40

4.

	A	B
	X	Y
1	4.3	7.31
2	5.1	8.67
3	6.5	11.05
4	7.0	11.90
5	9.6	16.32

5.

	A	B
	X	Y
1	2.3	2.76
2	2.6	3.64
3	2.9	4.64
4	3.2	5.76
5	3.5	7.70

6.

	A	B
	X	Y
1	9.6	6.72
2	8.1	5.67
3	7.3	5.11
4	5.1	3.57
5	2.9	2.03

7.

	A	B
	X	Y
1	−3.5	4.20
2	5.5	−6.60
3	2.1	−2.52
4	4.8	−5.76
5	3.6	−4.32

8.

	A	B
	X	Y
1	3.5	4.20
2	5.5	−6.60
3	2.1	−2.52
4	4.8	−5.76
5	3.6	4.32

9. Without using a spreadsheet, explain why the table in Exercise 8 cannot represent a direct-variation relationship between x and y.

Technology
1.5 Spreadsheets and Solving Equations

You can use a spreadsheet to help you solve an equation like $0.826x - 2.961 = 3.61x - 20.501$. Set up a spreadsheet so that column A contains a range of values for x, such as 0 to 10, column B contains the values of $0.826x - 2.961$, the left side of the equation, and column C contains the values of $3.61x - 20.501$, the right side of the equation. From the spreadsheet, you can observe that the values in columns B and C are closest when $x = 6$. Revise column A by using 5.6, 5.7, 5.8, 5.9, 6.0, 6.1, 6.2, 6.3, 6.4, 6.5 as shown. The values in columns B and C are closest when $x = 6.3$. Thus, the solution of $0.826x - 2.961 = 3.61x - 20.501$ is about 6.3, to the nearest tenth.

	A	B	C
1	X	LEFT	RIGHT
2	5.6	1.6646	−0.2850
3	5.7	1.7472	0.0760
4	5.8	1.8298	0.4370
5	5.9	1.9124	0.7980
6	6.0	1.9950	1.1590
7	6.1	2.0776	1.5200
8	6.2	2.1602	1.8810
9	6.3	2.2428	2.2420
10	6.4	2.3254	2.6030
11	6.5	2.4080	2.9640

Notice from the spreadsheet that the values in columns B and C are "far apart," then closer together, and then farther apart. This condition must be satisfied for the equation to have a solution between the least and greatest x-values selected.

Use a spreadsheet to approximate any solution in the given range to the nearest tenth. If no solution exists, state so.

1. $2.3x - 5 = 4x - 7.55$ for all x

   _____

2. $6x + 4 = 5x + 14.3$ for all x

   _____

3. $-2x + 6 = 3x + 14.5$ for all x

   _____

4. $1.3x - 4 = 2x - 0.5$ for all x

   _____

5. $2x - 1.9 = 3x - 5.9$ for $-4 \leq x \leq 0$

   _____

6. $6.2x + 9 = 4x + 22.2$ for $5 \leq x \leq 7$

   _____

7. $13.9x + 5 = 17.1x + 22.5$ for $-9 \leq x \leq 0$

   _____

8. $6.3x - 17 = -18x + 206.8$ for $0 \leq x \leq 20$

   _____

9. $33.6x + 17 = 93$ for $-9 \leq x \leq 0$

   _____

10. $100.37 = -19x - 206.8$ for all x

    _____

Technology
1.6 Graphics Calculators and Inequalities

You can use your knowledge of solving equations and a graphics
calculator to help you solve the inequality $3x - 1 > 2x + 3$.

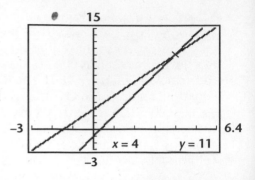

- Graph $y_1 = 3x - 5$ and $y_2 = 2x + 3$ on the same coordinate
  plane.
- Move the trace cursor to the point where the lines intersect and
  read the x-coordinate of that point.
- The line representing $y_1 = 3x - 1$ (the line with y-intercept -1
  is above the line representing $y_2 = 2x + 3$ (the line with
  y-intercept 3) when x is greater than 4.

So, the solution of $3x - 1 > 2x + 3$ is $x > 4$. Now, you can graph
$x > 4$ on a number line as shown.

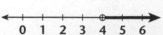

**Use a graphics calculator to solve each inequality. Then
graph the solution on a number line.**

1. $4x + 1 < 2x - 3$

   _____

2. $6.25x \geq 4.25x + 8$

   _____

3. $-2x + 5 \leq -2x + 3$

   _____

4. $-2x - 5 \leq -2x + 3$

   _____

5. $3x + 7 \geq 2(x + 6)$

   _____

6. $4x - 6 > 12$

   _____

7. $9 \leq 4x + 9$

   _____

8. $7x - 5 < 8x + 1$

   _____

9. $3x + 4 \geq 4$ and $2x - 6 \leq 0$

   _____

10. $3x - 1 \geq 3x - 1$

    _____

11. $4x + 6.2 \geq -2x + 160.5$

    _____

12. $5(x - 1) \geq -3(x - 2) + 5$

    _____

Lesson Activity
1.1 Human Head Hair

On average, human head hair grows at a rate of about 0.5 inches per month. The table shows this relationship over a period of a year. If a person's hair is not cut, it grows to a maximum of about 2 to 3 feet.

x	1	2	3	4	5	6	7	8	9	10	11	12
y	0.5	1.0	1.5	2.0	2.5	3.0	3.5	4.0	4.5	5.0	5.5	6.0

1. Describe what x and y represent.

_____

2. Are x and y linearly related? Explain. Include in your explanation whether you think the graph of this relationship is increasing or decreasing.

_____

_____

3. How would you use the given table to predict the length of your own hair after a period of time?

_____

4. If a person wants to let his or her hair grow to a maximum length of 2 to 3 feet, how long would it take?

_____

5. Describe how you determined your answer to Exercise 4.

_____

6. Measure the average hair length of several family members and record your data in the chart provided. Then determine how long it would take to double the length of each person's average hair length.

Name	Length of Hair	2 × length	Number of Months

Lesson Activity
1.2 Will There Be a Rain-Off?

The slope of a structure allows for rain water to run off. Some recommended minimum slopes to allow for proper drainage are listed.

Driveway $1\frac{1}{2}$ in. per ft   Gutter $\frac{1}{8}$ in. per ft

Skylight   3 in. per ft   Roof Deck $\frac{1}{4}$ in. per ft

Write equations to model the slope for proper drainage for each structure where the drop d, in inches, is equal to the product of the slope and the length l of the structure in feet.

1. driveway _____   2. gutter _____

3. skylight _____   4. roof deck _____

Graph each of the equations on the grid provided.

5. Driveway

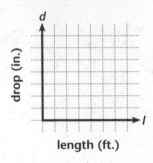

6. Gutter

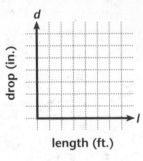

7. Skylight

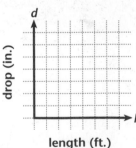

8. Roof Deck

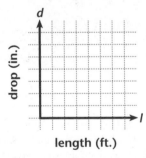

9. The recommended steepness for the incline of a handicap ramp is 1 inch of rise for every 10 feet of run. Write an equation to model the height (rise) of the ramp based on its length (run).

_____

10. Measure several handicap ramps and determine if they meet the recommended steepness. Record your data in a chart such as the one shown.

Ramp Location	Run (in in.)	Rise (in ft)	Incline	Meets $\frac{1}{10}$?

Lesson Activity
1.3 Recorded Music Sales

The table shows the percent of recorded music sales in CDs, tapes, and vinyl records.

Percent of Recorded Music Sales

Year	CD	Tape	Vinyl
1989	36%	50%	9%
1990	42%	48%	5%
1991	50%	43%	2%
1992	56%	37%	1%
1993	61%	33%	0.2%

Source: RIAA Consumer Profile

1. Use the grid provided to draw three scatter plots to represent the data. Use different symbols to represent CDs, tapes, and vinyl records.

Identify the correlation as positive, negative, or zero.

2. CD _____

3. tape _____

4. vinyl _____

Find the correlation coefficient *r*, for each recorded music medium.

5. CD _____ 6. tape _____ 7. vinyl _____

Write the equation of the line of best fit for each recorded music medium.

8. CD _____ 9. tape _____ 10. vinyl _____

11. Use the line of best fit for CDs to predict the percent of CD music sales in 2000. Is this realistic?

_____

12. Do the percent of CDs, tapes, and vinyl records add to 100%? Why or why not?

_____

Lesson Activity
1.4 Poster Proportions

For printing purposes, poster paper comes in standard sizes called B series as shown with dimensions in millimeters.

Dimensions (in mm)		Dimensions (in mm)	
B0	1000×1414	B6	125×176
B1	707×1000	B7	88×125
B2	500×707	B8	62×88
B3	353×500	B9	44×62
B4	250×353	B10	31×44
B5	176×250		

1. If inches vary directly with mm and 1 in. = 25.4 mm, find what 1 mm

   equals in inches. _____

Find the dimensions of each B series poster in inches.

2. B0 _____

3. B1 _____

4. B2 _____

5. B3 _____

6. B4 _____

7. B5 _____

8. B6 _____

9. B7 _____

10. B8 _____

11. B9 _____

12. B10 _____

13. Describe any patterns you observe.

   _____

14. Label the given diagram to show which of the B series poster sizes is represented by each of the rectangles.

15. Measure several posters you have at home or in your classroom. Identify any posters that are not represented by the B series.

   _____

   _____

   _____

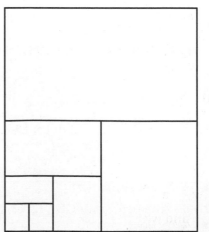

NAME _____ CLASS _____ DATE _____

Lesson Activity
1.5 Shirts for Sales

The school Spirit Club wants to raise money by selling
T-shirts with the school colors and mascot. Creative Shirts
charges a fixed fee of $75 for creating the design and $5 for
each shirt they print.

1. Write an equation for the cost of printing shirts with
   Creative Shirts.

   _____

2. Write an equation describing the Spirit Club's income
   if they sell each T-shirt for $8.

   _____

The point where the graphs of these two equations intersect is called the
break-even point. In this case, the break-even point describes the minimum
number of shirts the spirit club must print and sell so they do not lose
money.

3. What is this break-even point? _____

4. How does the break-even point change if the Spirit Club sells the
   T-shirts for $10?

   _____

5. How many shirts at $8 must they sell to make a $525 profit? _____

Another company, Custom Clothes, is approached about printing the
T-shirts. They charge a fixed fee of $105 per shirt design and $4.50 for each
shirt they print.

6. Write an equation for the cost of printing shirts with Custom Clothes.

   _____

7. If the Spirit Club sells the T-shirts for $8, how many shirts must they

   sell to break even using Custom Clothes? _____

8. How many shirts at $8 must they sell to make a $525 profit? _____

9. Which company would you recommend for printing the T-shirts

   and why? _____

10. For how many shirts are Creative Shirts and Custom Clothes equal options? _____

Lesson Activity
1.6 Fly a Kite

The kite has been formed by four linear inequalities.

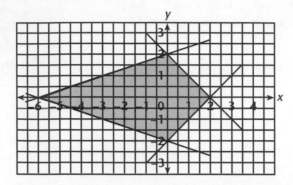

1. Write linear inequalities for each line that forms the sides of the kite.

   _____

   _____

2. Describe what you observe about the sides and diagonals of this kite.

   _____

   _____

3. Write an inequality that would form a "tail" for the kite. Be sure to
   make the tail a specific length.

   _____

4. Use the grid provided to draw any geometric figure and write linear
   inequalities for each of its sides.

   _____

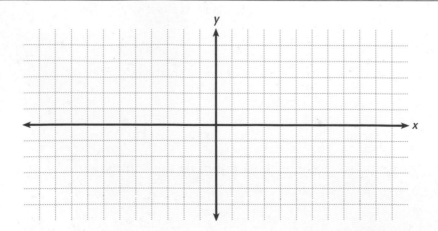

Assessing Prior Knowledge
1.1 Tables and Graphs of Linear Equations

1. Plot the orders pairs (2, 10), (−1, 4), and (5, 16) on the grid provided.

2. Connect the points.

3. Estimate the *y*-value when the *x*-value is zero.

   _____

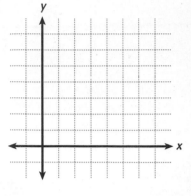

Quiz
1.1 Tables and Graphs of Linear Equations

Graph each equation on the grid provided.

1. $y = -x + 2$

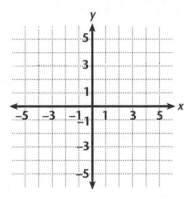

2. $2y = 4x + 6$

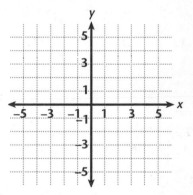

3. Which of the graphs above is increasing? Explain.

   _____

   _____

4. Jorge wants to join The Fitness Center. The monthly club fee is $15 plus $3 each time he swims in the club's pool. Write an equation to represent the monthly cost *c*, to belong to The Fitness Center in terms of pool usage *p*.

   _____

The Fitness Center
Membership Card

FEE: $15 plus
$3 per pool visit

Assessing Prior Knowledge
1.2 Exploring Slopes and Intercepts

1. Graph $y = 2x + 1$ by making a table of values.

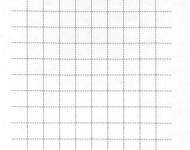

2. What is slope?

_____

_____

3. Will the slope change anywhere on a line? _____

- -

Quiz
1.2 Exploring Slopes and Intercepts

Identify the slope *m*, and the *y*-intercept *b*, of each line. (If the slope is undefined, or if there is no *y*-intercept, indicate this.)

1. $y = 10$ 　　　　　**2.** $y = 2x - 10$ 　　　　　**3.** $y = -15x$

_____　　　_____　　　_____

4. $x = -7$ 　　　　　**5.** $2y = 4x - 2$ 　　　　　**6.** $-x = y$

_____　　　_____　　　_____

7. What is the slope of a line that contains the points (1, 3) and (−2, 6)?

_____

8. Write the equation of the line in Exercise 7 in slope-intercept form.

_____

Write a linear equation with the indicated slope, *m*, and *y*-intercept, *b*.

9. $m = 5, b = 2.5$ 　　　**10.** $m = -1, b = 0$ 　　　**11.** $m = 0, b = -2$

_____　　　_____　　　_____

Assessing Prior Knowledge
1.3 Scatter Plots and Correlation

1. Describe the graph of the equation $y = 2x + 4$. Is the slope positive or negative?

   _____

2. From looking at a graph, how can you tell if the slope is positive or negative?

   _____

   _____

- -

Quiz
1.3 Scatter Plots and Correlation

Use the following table to answer Exercises 1–4.

x	2	8	4	14	20	6	12	16	18	10
y	30	24	27	16	10	27	22	15	25	21

1. Make a scatter plot to represent the given data and draw the line of best fit.

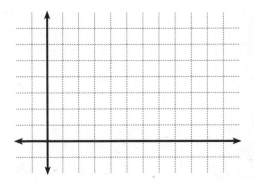

2. Does your scatter plot have a positive, negative, or zero correlation? _____

3. Explain what an outlier is and identify any outliers in the scatter plot you made in Exercise 1.

   _____

4. Using your line of best fit, predict the value of y when $x = 22$. _____

Mid-Chapter Assessment
Chapter 1 (Lessons 1.1 – 1.3)

Write the letter that best answers the question or completes the statement.

_____ **1.** Which of the following equations is *not linear*?

    a. $3y = 7x - 10$ **b.** $-2x - 3y = 12$
    c. $4y^2 = x$ **d.** $y = 2$

_____ **2.** The slope and *y*-intercept of the equation $2y = 2x + 10$ are, respectively

    a. 2, 10   **b.** 1, 5   **c.** 10, 2   **d.** 5, 1

_____ **3.** The slope of the line containing the points (2,4) and (2,1) is

    a. 0   **b.** 3   **c.** undefined   **d.** −3

_____ **4.** The equation of the line that has a slope of 3 and contains the point (1,3) is

    a. $x = 3y - 8$ **b.** $y = 3x$ **c.** $y = 3x + 3$ **d.** $y = 3$

Graph the following equations on your own graph paper.

5. $y = -x - 3$ **6.** $x = 4$

7. Jolene invests in baseball cards. One card she purchased for $20 increases in value by $5 each year. Write an equation that represents the value of the card *v* at the end of *n* years.

_____

8. The scatter plot displays the number of pounds of weight lost verses the number of hours of exercise performed per week by 18 participants. Is the correlation between weight loss and exercise positive, negative, or nonexistent?

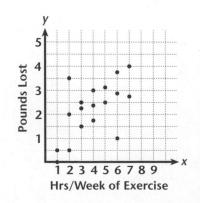

_____

9. List any outliers in the scatter plot.

_____

10. Estimate the correlation coefficient for the scatter plot. _____

Assessing Prior Knowledge
1.4 Direct Variation and Proportion

1. Using the equation $y = 4x$, complete the following ordered pairs: (2, ?), (3, ?), and (4, ?). _____

2. How are the x-values and the y-values related in each ordered pair?

    _____

3. Solve for x: $\frac{x}{2} = \frac{6}{8}$. _____

- -

Quiz
1.4 Direct Variation and Proportion

For each of the values of d and t, d varies directly as t. Find the constant of variation k and write the equation of direct variation.

1. d is 120.5 when t is 2.5

    _____

2. d is 15 when t is 0.05

    _____

3. d is $\frac{1}{4}$ when t is $\frac{2}{5}$

    _____

4. d is -45 when t is 0.9

    _____

In Exercises 5 and 6, c varies directly as d. Find the value for the indicated variable.

5. c is -2 when d is -24.4.
   Find c when d is 1.22.

    _____

6. c is 9 when d is 54.
   Find d when c is 4.

    _____

7. The number of shoes s on the gym floor varies directly as the number of basketball players p. Write an equation of direct variation that relates the two variables.

    _____

8. A car travels at 55 mi/h for 137.5 miles. How long will the trip take? _____

NAME _____ CLASS _____ DATE _____

Assessing Prior Knowledge
1.5 Solving Equations

1. Solve for x: $12 = 5x + 2$. _____

2. Use the Distributive Property to simplify $5(9x - 6)$. _____

- -

NAME _____ CLASS _____ DATE _____

Quiz
1.5 Solving Equations

Solve the following equations by graphing each on the grid provided.

1. $-3x + 2 = -1$

2. $\frac{1}{2}x + 2 = x + \frac{1}{2}$

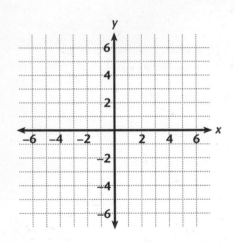

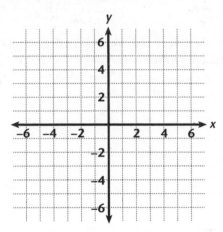

3. Given that $y = 4x - 1$, use substitution to solve $2y - 3x = 8$ for x.

_____

4. Solve $\frac{1}{3}\pi rh = V$ for r. _____

5. Solve $y = mx + b$ for m. _____

Solve the equation and show each step.

6. $14 - 2(x + 3) = 2x$

_____

_____

NAME _____ CLASS _____ DATE _____

 Assessing Prior Knowledge
1.6 Solving Inequalities

1. Write $x < 5$ in words. _____

2. Graph $x < 2$ on the number line.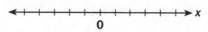
                                            0

– –

NAME _____ CLASS _____ DATE _____

 Quiz
1.6 Solving Inequalities

Graph the solution of each inequality on the number lines provided.

1. $\dfrac{10z}{-5} < z + 6$

2. $4x + 2 \geq -18 + 6x$

Graph the solution of each inequality on the grid provided.

3. $-3x - 1 \leq 2y$

4. $2x > \dfrac{2y - 3}{4}$

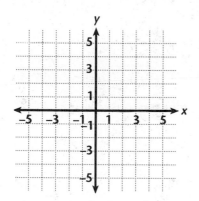   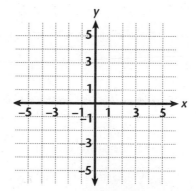

5. Write the inequality whose graph is shown.

_____

_____

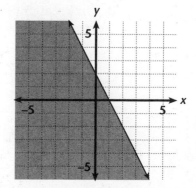

Chapter Assessment
Chapter 1, Form A, page 1

Write the letter that best answers the question or completes the statement.

_____ **1.** Which of the following equations represents a line?

 a. $x = 2$ **b.** $y^2 = 2x$ **c.** $y = x^3$ **d.** $2y = 3x^2 - x$

_____ **2.** The solution for the equation $2x + 3(x - 7) = -2(x - 21)$ is

 a. $x = 2$ **b.** $x = 9$ **c.** $x = 5$ **d.** $x = -3$

_____ **3.** The equation $a = kb$ represents a direct variation between a and b. Given that b is 2.1 when a is 21, the constant of variation k is

 a. 1   **b.** 0.1   **c.** 10   **d.** 100

_____ **4.** In which table are the variables linearly related?

 a.

x	0	3	6	9
y	0	6	12	18

 b.

x	−1	−2	−3	−4
y	1	4	9	16

 c.

x	−2	−1	0	1
y	−8	−1	0	1

 d.

x	0	1	2	3
y	1	2	5	10

_____ **5.** The slope of the line containing the points $(-2, 8)$, $(-1, 4)$, and $(2, -8)$ is

 a. 4   **b.** 0   **c.** $\frac{1}{4}$ **d.** −4

_____ **6.** The solution set for the inequality $-x - 3(x - 2) < -10 + 4x$ is

 a. $x \geq 2$ **b.** $x > 2$ **c.** $x < 2$ **d.** $x < -2$

_____ **7.** Trees-R-Us tree service charges a basic fee of $50 to make a house call plus $20 an hour to trim trees. The equation that represents the total cost c for Trees-R-Us to make a house call and trim trees in terms of the number of hours h is

 a. $c = 20h + 50$ **b.** $c = 50h + 20$ **c.** $c = 20h$ **d.** $c = 50h$

_____ **8.** In a scatter plot where no slope or line of best fit is evident, the correlation is

 a. positive   **b.** negative
 c. zero   **d.** none of the above

Chapter Assessment
Chapter 1, Form A, page 2

_____ 9. The y-intercept of a line having a slope of -2 and containing the point $(-2, 1)$ is

   a. 0   **b.** -5 **c.** 3   **d.** -3

_____ 10. When the equations $y = 7x - 1$ and $2y + 3x = 49$ are solved using substitution, the value of x is

   a. $2\frac{16}{17}$ **b.** 3   **c.** 5   **d.** -3

_____ 11. The following graph represents the solution set to which inequality?

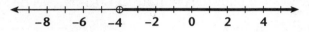

   a. $-3x - 1 < 11$ **b.** $-3x - 1 \leq 11$
   c. $-3x - 1 \geq 11$ **d.** $-3x - 1 > 11$

_____ 12. Which property is exemplified by $9(x - 1) = 9x - 9$.

   a. Addition Property   **b.** Substitution Property
   c. Division property   **d.** Distributive Property

_____ 13. The number of tires t on a street varies directly as the number of cars c on the street. The equation of direct variation for these two variables is

   a. $4c = t$ **b.** $c = 4 + t$
   c. $c = 4t$ **d.** none of the above

_____ 14. When $x = -2$, $y = 0.5$. Find y when $x = -40$.

   a. -10 **b.** 0.1   **c.** 10   **d.** 100

_____ 15. When solving $V = \frac{1}{4}l + w$ for l, l is equal to

   a. $4w$ **b.** $4V - w$ **c.** $V - w$ **d.** $4V - 4w$

_____ 16. The graph represents which of the following inequalities?

   a. $2y \geq x - 2$ **b.** $y \geq -\frac{1}{2}x - 2$

   c. $y \leq -2x - 2$ **d.** $y \leq \frac{1}{2}x - 2$

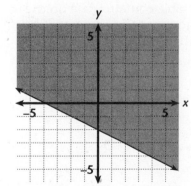

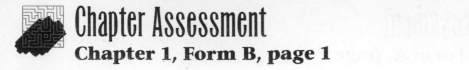

Chapter Assessment
Chapter 1, Form B, page 1

1. What is the slope and y-intercept of the equation $y + 6 = \frac{1}{3}x$?

_____

2. Graph the equation for Exercise 1 using your own graph paper.

3. Give one example of a linear equation and one example of a nonlinear equation.

_____

4. Write the equation of the line that contains the points $(5, -17)$, $(0, -2)$, and $(-3, 7)$.

_____

5. Find the equation of the line that contains the point $(2, -3)$ and has a slope of 1.

_____

Describe a scatter plot that would represent each situation.

6. $r = -1$ _____

7. $r = 0$ _____

Identify whether you would expect a positive correlation, a negative correlation, or no correlation for each pair of variables.

8. Number of accessories on a car and car price _____

9. Color of house and square footage of house _____

The table contains data on the number of car accidents within a 24-hour period compared with the amount of snow accumulated for the town of Rocky Hills.

Inches of snow	0	1.5	4	5	0.5	7	12	4	1.5	2.5	5.5	5.5	9	3	4.5
Accidents		2	9	6	5	1	10	15	0	5	2	6	7	11	4	8

Chapter Assessment
Chapter 1, Form B, page 2

10. Use the grid provided to make a scatter plot for the given data and draw the line of best fit.

11. List any outliers for the scatter plot in Exercise 10.

_____

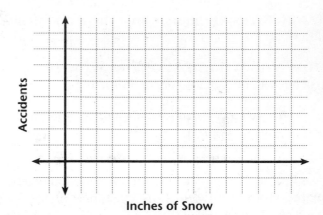

12. Is the correlation of the scatter plot in Exercise 10 positive or negative?

_____

In Exercises 13 and 14, v varies directly as t and $v = 14$ when $t = -5$.

13. Find the constant variation of k. _____

14. Find t when $v = -63$. _____

The amount of money Aaron makes, p, is directly proportional to the number of boats he sells, b. Aaron makes $1050 selling 3 boats.

15. Write an equation of direct variation in terms of p and b. _____

16. How much does Aaron make for selling 20 boats? _____

17. Solve the linear equation $19x + 3 + 14x = 6(x + 2)$. _____

18. Solve the equation $x = -3x + 6$ by graphing. Use your own graph paper.

In Exercises 19 and 20, graph the solution on the number line provided.

19. $5x + 22 \geq 7$ **20.** $14x > 2x - 24$

     ←—+—+—+—+—+—+—+—+—+—+—+—+—+—+—+→   ←—+—+—+—+—+—+—+—+—+—+—+—+—+—+—+→

In Exercises 21 and 22, use your own graph paper to graph each inequality.

21. $7x + 4y \geq 8$ **22.** $9y < -18x$

23. Given that $y = 4 - 2x$ and $23 = 3y - 13$, use substitution to solve for x. _____

NAME _____ CLASS _____ DATE _____

Alternative Assessment
Recognizing Linear Relationships, Chapter 1, Form A

TASK: To identify linear relationships and write the linear equation described

HOW YOU WILL BE SCORED: As you work through the task, your teacher will be looking for the following:

- how well you can identify a linear relationship between variables in a table
- whether you can write an equation describing a linear relationship
- how well you can identify a direct variation from a table

The junior class at Oakland High is conducting a fund raising activity by recycling aluminum, bi-metal, and/or tin cans. The class gets a $3.50 refund for 100 cans.

1. Assign variables to the number of cans recycled and the amount of refund.

_____

2. Make a table for the number of cans recycled and the amount of refund. Use consecutive values.

3. Describe how you can tell if the variables are linearly related.

_____

4. Write an equation that represents this relationship.

_____

5. How much will the class get for 500 cans? How many cans does the class need to recycle to raise $525?

_____

6. Is this linear equation a direct variation? Explain.

_____

SELF-ASSESSMENT: Write a direct variation equation describing a real-world proportional relationship.

---

36 **Alternative Assessment**   **HRW Advanced Algebra**

NAME _____ CLASS _____ DATE _____

Alternative Assessment
Solving Inequalities, Chapter 1, Form B

TASK: To solve inequalities that model real-world applications

HOW YOU WILL BE SCORED: As you work through the task, your teacher will be looking for the following:

- how well you can write an inequality describing a real-world application
- whether you can graph an inequality in one variable

Adventure software costs $15 for each game and sports games cost $10 each. If Demane has $120 to spend, what combinations of adventure software and sports games can Demane choose?

1. Assign variables to adventure software and sports games. Then write an inequality that describes the situation.

   _____

   _____

2. Solve the inequality you wrote for one of the variables.

   _____

3. Describe the values that are possible for this situation.

   _____

4. Sketch the graph showing how many of each games Demane could buy.

5. What will be the cost of 6 adventure software games and 4 sports games?

   _____

6. What combinations of adventure software and sports games can Demane purchase, if he wants to spend as much as possible of the $120.

   _____

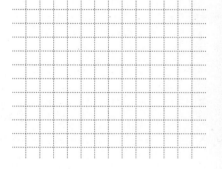

SELF-ASSESSMENT: What is the difference between real-world applications modeled by an equality and real-world applications modeled by an inequality?

Practice & Apply
2.1 Exploring Numbers

Simplify.

1. $\dfrac{3}{7} + \dfrac{1}{4}$ _____

2. $\dfrac{2}{3} \cdot \dfrac{9}{14}$ _____

3. $\dfrac{x}{2} - \dfrac{5x}{4}$ _____

4. $\dfrac{a^2}{2c} \div \dfrac{a}{4c}$ _____

5. $\dfrac{8}{y-5} \cdot \dfrac{y-5}{12}$ _____

6. $\dfrac{n}{5} + \dfrac{1+3n}{3}$ _____

7. $\dfrac{2x-6}{6} \div \dfrac{(x+3)(x-3)}{4}$ _____

8. $\dfrac{3(y-1)}{3} - \dfrac{5y}{6}$ _____

9. $\dfrac{(y+5)(y-5)}{y+5} \cdot \dfrac{y}{5y-25}$ _____

10. $\dfrac{-2(1-a)}{5} + \dfrac{3a}{4}$ _____

11. $\dfrac{2+3c}{9} - \dfrac{3c}{2}$ _____

12. $\dfrac{d}{6c} \div \dfrac{9d}{8c^2}$ _____

Let *a*, *b*, and *c* represent any real numbers. Name the property illustrated.

13. $a + b = b + a$ _____

14. $(a \cdot b) \cdot c = a \cdot (b \cdot c)$ _____

15. $a \cdot (b + c) = a \cdot b + a \cdot c$ _____

16. $\dfrac{1}{a} \cdot a = 1$ _____

17. The length of a rectangular garden is $\dfrac{5y}{2y-4}$ meters and its width is $\dfrac{(y+2)(y-2)}{10}$ meters. Find the area of the garden.

Practice & Apply

2.2 Exploring Properties of Exponents

Simplify.

1. $x^4 \cdot x^3$ _____

2. $\dfrac{d^8}{d^2}$ _____

3. $(y^{-3})(y^5)$ _____

4. $(5x^2y)(4xy^3)$ _____

5. $\dfrac{-24ab^{-2}}{6a^{-1}b^{-3}}$ _____

6. $(3^{\frac{1}{2}})(3^2)$ _____

7. $(2n^3)^4$ _____

8. $(3x^{-2}y)(4xy^{-4})$ _____

9. $\left(\dfrac{2y^{-2}}{3y}\right)^2$ _____

10. $(7^2)(7^{3.2})$ _____

11. $(3c^{-2})^{-3}$ _____

12. $(3^9)^{\frac{4}{9}}$ _____

13. A storage facility has the shape of a cube. The length of each side is y feet. Use exponents to write the formula for volume of the storage facility.

_____

14. Graph the equations $y = \dfrac{x^5}{x^2}$ and $y = x^3$ on the same coordinate plane.

Make a table of values for each equation using the x-values -2, -1, 0, 1, and 2. How are the graphs and tabular values the same? How are they different? Explain your results using the rules of exponents you have developed.

_____

_____

_____

Practice & Apply
2.3 Definition of a Function

Determine whether the set of ordered pairs represents a function.

1. {(1, 3), (2, 4), (3, 5), (4, 6)}

_____

2. {(0, 4), (2, 1), (2, 7), (5, −3)}

_____

Determine whether each table represents a function.

3.

x	y
1	2
2	3
3	4
4	5

_____

4.

x	y
1	7
3	7
5	7
7	9

_____

5.

x	y
1	0
3	−1
3	−2
8	−3

_____

6.

x	y
2	−8
4	−6
6	−4
8	−2

_____

7.

x	y
1	0
1	−1
2	5
2	−4

_____

Determine whether each graphed relation is a function.

8.

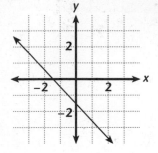

_____

9.

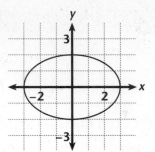

_____

10.

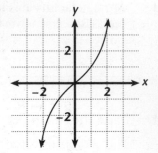

_____

Use your graphics calculator to graph each relation. Find the domain and range of each. Determine whether the relation is a function.

11. $y = x - 1$ _____

12. $y = x^2 + 3$ _____

13. $x = y^2$ _____

14. $y = \frac{1}{x}$ _____

An electronics store is offering 25% off on all VCRs.

15. Write an equation in which the discounted price is a function of the original price. _____

16. Use your graphics calculator to graph this function.

17. Gilbert purchased a VCR that originally sold for $125. Use the trace feature on your calculator to determine the sale price of the VCR. _____

Practice & Apply
2.4 Using Notation to Represent Functions

In Exercises 1–12, let $f(x) = 5x - 3$, $g(x) = x^2 - 4x - 6$, and $h(x) = \dfrac{2}{x} - 3$.

1. $f(6)$ _____

2. $g(-1)$ _____

3. $h(5)$ _____

4. $f(a)$ _____

5. $g(3)$ _____

6. $h(-7)$ _____

7. $f(x + h)$ _____

8. $g(x + 1)$ _____

9. $h(2x - 1)$ _____

10. $f(3r - 2)$ _____

11. $g(2n - 3)$ _____

12. $\dfrac{f(a) - f(b)}{a - b}$ _____

In Geometry, the circumference of a circle is π times its diameter.

13. Express the circumference of a circle using function notation.

_____

14. Find the circumference if the diameter is $3x$. _____

For Exercises 15–18, let $f(x) = -x^2 + 2x + 3$.

15. Use your graphics calculator to graph f.

16. Find the domain of f. _____

17. Find the vertex of f. _____

18. Find the range of f. _____

For Exercises 19–20, write an equation that specifies the function.

19. The cost $C(x)$ for renting a car is $25 per day plus 20¢ per mile for x miles. Express this relationship using function notation.

_____

20. Find the cost of renting a car for two days for a total of 250 miles.

_____

Practice & Apply
2.5 Linear Functions

Write the linear function f using the given information.

1. $m = 5, b = 7$ _____

2. $m = \frac{1}{3}, b = -1$ _____

3. $m = -4, f(0) = 2$ _____

4. $m = 1, f(0) = 6$ _____

Find the slope of the linear function containing the given points, and write the function.

5. $(-1, 5), (-2, 3)$ _____

6. $(-3, 5), (1, 2)$ _____

7. $(0, -2), (-3, -4)$ _____

8. $(2, 5), (5, 2)$ _____

Determine the linear function containing the points shown.

9. $(2, -5), (-3, 5)$ _____

10. $(-2, 4), (-6, 1)$ _____

11. $(3, 1), (2, -4)$ _____

12. $(4, 3), (-2, 6)$ _____

Determine whether the linear function containing these points is increasing, decreasing, or neither.

13. $(-2, 5), (3, 0)$ _____

14. $(-6, -3), (4, 1)$ _____

15. $(-2, 1), (5, 3)$ _____

16. $(4, 3), (6, -1)$ _____

Write the linear function of each graph.

17.

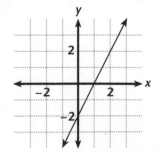

_____

18.

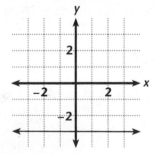

_____

19.

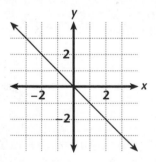

_____

Use your graphics calculator to graph each function. Find the x-intercepts.

20. $f(x) = 3x - 6$ _____

21. $f(x) = -2x + 5$ _____

22. $f(x) = \frac{x + 3}{4}$ _____

23. $f(x) = x^2 + 3$ _____

 **Practice & Apply**
2.6 Exploring Operations With Functions

Find $(f + g)(x)$ and $(f - g)(x)$.

1. $f(x) = -3x + 1; g(x) = x^2$ _____

2. $f(x) = x^2 - 2; g(x) = -3x + 7$ _____

3. $f(x) = \frac{1}{x}; g(x) = 4x$ _____

4. $f(x) = \sqrt{x}; g(x) = 5\sqrt{x}$ _____

5. $f(x) = \frac{2x}{x - 3}; g(x) = 2x$ _____

6. $f(x) = \frac{-x}{x + 2}; g(x) = 3$ _____

Find $(f \cdot g)(x)$ and $(f \div g)(x)$.

7. $f(x) = 6x; g(x) = 2x$ _____

8. $f(x) = 4x - 11; g(x) = x - 1$ _____

9. $f(x) = \frac{1}{2}x; g(x) = 4x - 8$ _____

10. $f(x) = x; g(x) = x - 7$ _____

11. $f(x) = \sqrt{3x}; g(x) = -x$ _____

12. $f(x) = \sqrt{x^2 - 16}; g(x) = x + 4$ _____

Let $f(x) = x^2 + 4$ and $g(x) = 2x - 1$. Use your graphics calculator to graph each function. Find the domain and range of each function.

13. f _____   **14.** $f + g$ _____   **15.** $f \cdot g$ _____   **16.** $f \cdot \frac{1}{g}$ _____

17. Let $f(x) = 3x - 2$. Use your graphics calculator to graph f and $-f$ on the same coordinate plane. What is the point of intersection? Compare the coordinates of f and $-f$.

_____

18. Linda's weekly earnings from selling cosmetics are represented by the function $E(x) = 250 + 0.04x$, where x is the dollar value of sales. Her sister's earnings from selling shoes are represented by the function $S(x) = 225 + 0.03x$. Find $(E + S)(x)$ and determine their combined income if they each make sales of $2000.

_____

Enrichment
2.1 A Different Kind of Rational Number

A rational number whose numerator is 1 is called a unit fraction. The ancient Egyptians did not write rational numbers as we do. If the numerator of a rational number was not a 1, they would write the rational number as the sum of a series of unit fractions. In addition, each denominator in the sum had to be a different number. For example, the ancient Egyptians would write the fraction $\frac{3}{4}$ as the sum of the unit fractions $\frac{1}{2}$ and $\frac{1}{4}$. That is, $\frac{3}{4} = \frac{1}{2} + \frac{1}{4}$.

Write each given rational number as a sum of unit fractions. You can use as many fractions as you need in any given sum as long as the denominators are different.

1. $\frac{3}{8}$ _____

2. $\frac{9}{10}$ _____

3. $\frac{4}{15}$ _____

4. $\frac{11}{12}$ _____

5. $\frac{4}{27}$ _____

6. $\frac{83}{90}$ _____

7. $\frac{43}{80}$ _____

8. $\frac{11}{18}$ _____

9. $\frac{51}{52}$ _____

10. $\frac{26}{75}$ _____

11. $\frac{66}{91}$ _____

12. $\frac{31}{40}$ _____

13. $\frac{5}{12}$ _____

14. $\frac{17}{20}$ _____

15. $\frac{8}{35}$ _____

16. $\frac{7}{15}$ _____

17. $\frac{8}{25}$ _____

18. $\frac{19}{26}$ _____

HRW Advanced Algebra

Enrichment
2.2 Compound Interest

Some interest is compounded more than once a year. To compute the amount A, of money after t years, use the formula

$$A = P\left(1 + \frac{r}{n}\right)^{nt}.$$

Here, P is the principal, the amount originally deposited, r is the rate of interest in decimal form, and n is the number of pay periods per year. For example, if $1500 is deposited at an annual interest rate of 6% compounded quarterly, then at the end of 3 years,

$$A = 1500\left(1 + \frac{0.06}{4}\right)^{4 \cdot 3}$$

$$= 1500\,(1.015)^{12}$$

$$\approx \$1793.43$$

Use a calculator to find the total amount of money after the given number of years at the given annual interest rates.

1. $100, 5 years, 5% interest compounded semiannually _____

2. $200, 4 years, 6% interest compounded semiannually _____

3. $500, 6 years, 4% interest compounded semiannually _____

4. $500, 6 years, 4% interest compounded quarterly _____

5. $1000, 8 years, 6% interest compounded quarterly _____

6. $1200, 10 years, 5% interest compounded quarterly _____

7. $100, 12 years, 6% interest compounded quarterly _____

8. $2000, 15 years, $5\frac{1}{4}$% interest compounded quarterly _____

9. $2599, 9 years, $7\frac{1}{2}$% interest compounded monthly _____

10. $3000, 20 years, $6\frac{1}{4}$% interest compounded monthly _____

11. $100, 25 years, 4% interest compounded semiannually _____

12. $100, 25 years, 4% interest compounded quarterly _____

Enrichment
2.3 Function Maze

**Find your way through this maze by moving from a square
containing a function to another square containing a function.
You may move vertically or horizontally one square at a time.**

START

$y = 2x$	$x = y^2$	$x^2 + y^2 = 1$	$\frac{x^2}{4} - \frac{y^2}{9} = 1$	$5x^2 - 7y^2 = 4$	$5x^2 + 5y^2 = 1$
$x + 3y = 4$	$x = y^2 + 3y$	$y = x^3$	$y = x + 4$	$y = x^2 - x$	$5x = 3y^2$
$y = \sqrt{x}$	$x^2 - y^2 = 49$	$y = x^2 + 3x + 1$	$2x^2 + 3y^2 = 4$	$x = 2y + 1$	$x + 2y^2 = 6$
$x = 2y$	$y = \frac{1}{x}$	$y = x^2$	$x = \frac{1}{y^2}$	$y - 3 = 4(x + 2)$	$x = 4$
$x = -2$	$x = \sqrt{7}y^2$	$2x^2 + 2y^2 = 3$	$15x^2 - 4y^2 = 1$	$y = \sqrt{x + 2}$	$x^2 + y^2 = 36$
$y + x = x^2$	$y = \frac{3}{2}x + 7$	$y = \frac{x^2 - 4}{2}$	$y = x^4$	$y = \frac{1}{x + 1}$	$5y^2 = x^2$
$4x - 3y = 8$	$\frac{x}{y^2} = 4$	$x = 10$	$x = \frac{y^2}{9}$	$\frac{x}{4} = y^2$	$2x^2 - 5y^2 = 8$
$y = 4x - 7$	$y = x^3 - 3x$	$y = x^5 + 2$	$y = 3x^2$	$y = \frac{x^2}{4}$	$y = 6$

EXIT

Enrichment
2.4 Tangents

Notice that l_1 intersects the graph of the parabola in two points, whereas l_2 intersects the graph in only one point. l_2 is called a *tangent line*—a line that touches a graph at only one point without crossing the graph. In calculus, the expression

$$\frac{f(x + h) - f(x)}{(x + h) - x}$$

can be used to find the slope of the tangent line at any point on the graph of the function f, by evaluating what happens when h gets smaller. That is, when h approaches zero.

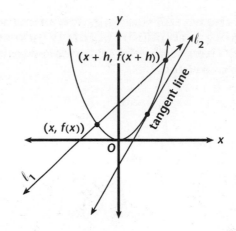

For example, to find the slope of the tangent line at any point on the graph of the function $f(x) = x^2 + 1$:

Evaluate.
$$\frac{f(x + h) - f(x)}{(x + h) - x} = \frac{[(x + h)^2 + 1] - [x^2 + 1]}{(x + h) - x}$$

Simplify.
$$= \frac{[x^2 + 2xh + h^2 + 1] - [x^2 + 1]}{h}$$

Combine like terms.
$$= \frac{2xh + h^2}{h}$$

Factor and simplify.
$$= \frac{h(2x + h)}{h}$$

$$= 2x + h$$

As h becomes very small, or approaches 0, the expression for the slope of the tangent at any x-value on the graph of the function f is $2x$.

Find the slope of the tangent line for each function at any x-value.

1. $f(x) = 3x + 8$

2. $f(x) = \frac{3}{4}x + 1$

3. $f(x) = -2x + 7$

4. $f(x) = x^2$

5. $f(x) = -x^2$

6. $f(x) = x^2 + 7$

7. $f(x) = x^2 - 1$

8. $f(x) = 2x^2 + 3x$

9. $f(x) = -3x^2 + 4x$

Enrichment
2.5 Increasing, Decreasing, or Neither

To find the mystery saying, determine whether the line containing each pair of points is increasing, decreasing, or neither. Then write the letter corresponding to your answer on the line provided.

_____ **1.** (2, 7) and (5, 6)
        I: M   D: S   N: T

_____ **2.** (−17, −7) and (−3, 10)
        I: L   D: A   N: E

_____ **3.** (2, 6) and (4, 6)
        I: T   D: R   N: O

_____ **4.** (5, 16) and (8, 14)
        I: N   D: P   N: R

_____ **5.** (−3, 7) and (2, 18)
        I: I   D: E   N: R

_____ **6.** (4, −3) and (5, −8)
        I: B   D: N   N: R

_____ **7.** (9, −4) and (−1, −4)
        I: O   D: K   N: G

_____ **8.** (7, 9) and (4, 10)
        I: T   D: L   N: A

_____ **9.** (5, 0) and (4, 2)
        I: M   D: I   N: N

_____ **10.** (−5, 3) and (−8, −1)
        I: N   D: J   N: Y

_____ **11.** (7, 15) and (8, 12)
        I: P   D: E   N: W

_____ **12.** (4, −2) and (8, 9)
        I: S   D: R   N: T

_____ **13.** (26, 19) and (17, 26)
        I: E   D: C   N: N

_____ **14.** (14, 17) and (18, 17)
        I: R   D: T   N: A

_____ **15.** (21, 14) and (25, −14)
        I: O   D: N   N: R

_____ **16.** (−14, 26) and (−18, 27)
        I: H   D: S   N: T

_____ **17.** (−2, 6) and (−1, 4)
        I: E   D: L   N: N

_____ **18.** (4, 7) and (7, 4)
        I: C   D: A   N: R

_____ **19.** (6, −3) and (9, −4)
        I: M   D: N   N: T

_____ **20.** (8, 15) and (6, 17)
        I: O   D: T   N: P

21. What phrase is spelled out by your answers?

_____

Enrichment
2.6 Multiplication Maze

**Find your way through the maze by moving only to a square
where $(f \cdot g)(x)$ is shown for each $f(x)$ and $g(x)$. You may move one
square horizontally or vertically at a time.**

ENTER

$3x + 13$	$3x - 14$	$x + 1$	$x + 5 + \frac{3}{x}$	$x^2 + x$
$f(x) = 2x + 7$	$f(x) = 4x - 6$	$f(x) = x^2 + 2x + 1$	$f(x) = x^2 + 5x + 3$	$f(x) = x$
$g(x) = x + 6$	$g(x) = x + 8$	$g(x) = x + 1$	$g(x) = x$	$g(x) = x + 1$
$x^2 + 10x + 3$	$2x - 1 + \frac{3}{x}$	$-14x^2 - 7x$	$24x$	$\frac{x}{2}$
$f(x) = 2x^2 + 7x + 4$	$f(x) = 4x^2 - 2x + 6$	$f(x) = 2x + 1$	$f(x) = 3x$	$f(x) = 4x$
$g(x) = x^2 - 3x + 1$	$g(x) = 2x$	$g(x) = -7x$	$g(x) = 8$	$g(x) = \frac{1}{8}$
$x^3 - x^2 + x$	$x^2 + 2x + 6$	$2x^2 - 9x - 5$	$16x^2 + 48x - 6$	$2x^2 + 2x - 6$
$f(x) = x^2 - x + 1$	$f(x) = 2x$	$f(x) = 2x + 1$	$f(x) = 2x + 6$	$f(x) = x^2 + 2x$
$g(x) = \frac{1}{x}$	$g(x) = x^2 + 6$	$g(x) = x - 5$	$g(x) = 8x - 1$	$g(x) = x^2 - 6$
$x - \frac{2}{x}$	$x + 2$	$-x^3 + x^2$	$x^2 + 3x - 18$	$x^3 - 2x^2 + x - 2$
$f(x) = x^2$	$f(x) = x^2 - 3x + 4$	$f(x) = x^2$	$f(x) = x + 6$	$f(x) = x - 2$
$g(x) = x + 2$	$g(x) = x^2 - 2x + 6$	$g(x) = -x + 1$	$g(x) = x - 3$	$g(x) = x^2 + 1$
$\frac{x + 2}{x - 1}$	$2x$	$x^2 + 7x - 8$	$16x^2 + 46x + 5$	$-14x^3$
$f(x) = x^2 + 5x + 6$	$f(x) = \frac{2}{x}$	$f(x) = 3x - 8$	$f(x) = 2x + 6$	$f(x) = -7x$
$g(x) = x^2 + 2x - 3$	$g(x) = \frac{1}{x^2}$	$g(x) = x^2 + 4x$	$g(x) = 8x - 1$	$g(x) = 2x^2$
$x^2 + x$	$12x^2 - 23x + 10$	$x + 2 - \frac{6}{x}$	$\frac{x - 7}{x}$	1
$f(x) = \frac{1}{x}$	$f(x) = 2 - 3x$	$f(x) = x^2 + 2x - 6$	$f(x) = x - 7$	$f(x) = \frac{1}{x + 2}$
$g(x) = x + 1$	$g(x) = 5 - 4x$	$g(x) = \frac{1}{x}$	$g(x) = \frac{1}{x}$	$g(x) = x + 2$

EXIT

HRW Advanced Algebra   **Enrichment**   **49**

Technology
2.1 Approximating Square Roots

You already know that $\sqrt{51}$ is a real but not rational number. This means that $\sqrt{51}$ cannot be written as a terminating or repeating decimal. It can, however, be approximated by a rational number. To carry out the approximation process, consider the following logic. Since $7^2 < \sqrt{51}^2 < 8^2$, 7 is the largest whole number less than $\sqrt{51}$. So, 7 is a first approximation to $\sqrt{51}$.

Now find the average of 7 and $\frac{51}{7}$.

$$\frac{7 + \frac{51}{7}}{2} = 7.14285714$$

That number is the second approximation to $\sqrt{51}$. By continuing the process suggested by the logic above, you can get a better and better approximation of $\sqrt{51}$. The spreadsheet shows how the thinking above can be carried out on a computer.

	A	B	C
1	APPROXIMATION	QUOTIENT	AVERAGE
2	7.00000000	7.28571429	7.14285714
3	7.14285714	7.14000000	7.14142857
4	7.14142857	7.14142829	7.14142843
5	7.14142843	7.14142843	7.14142843
6	7.14142843	7.14142843	7.14142843
7	7.14142843	7.14142843	7.14142843

Cell A2 contains the largest integer less than $\sqrt{51}$.
Cell B2 contains 51/A2.
Cell C2 contains (A2+B2)/2.
Cell A3 contains C2.
Cell B3 contains 51/A3.
Cell C3 contains (A3+B3)/2.

Cell A7 contains the sixth approximation to $\sqrt{51}$. That approximation is 7.14142843.

Use a spreadsheet like the one above to approximate the value of each expression.

1. $\sqrt{2}$ **2.** $\sqrt{5}$ **3.** $\sqrt{18}$ **4.** $\sqrt{96}$

_____ _____ _____ _____

5. $\sqrt{101}$ **6.** $\sqrt{300}$ **7.** $\sqrt{50}$ **8.** $-\sqrt{12}$

_____ _____ _____ _____

9. $\sqrt{3} - \sqrt{2}$ **10.** $\sqrt{12} + \sqrt{24}$

_____ _____

Technology

2.2 Compound Interest Tables

When you deposit money into a bank and interest is paid on all money present in the account, you receive compound interest. The following formula gives the amount A in the account after t years given the listed data.

$$A = P\left(1 + \frac{r}{n}\right)^{(nt)} \begin{cases} P = \text{amount of deposit} \\ r = \text{yearly interest rate as a decimal} \\ n = \text{the number of times per year interest is calculated} \end{cases}$$

If, for example, $P = \$750$, $r = 8.5\%$, and interest is calculated annually (that is, $n = 1$), then you get the formula $A = 750(1.085)^t$.

The spreadsheet gives the amount in the account over the first three years given $P = \$750$, $r = 8.5\%$, and $n = 1$.

	A	B	C	D	E
1	P	R	N	T	A
2	750	0.085	1	0	750.00
3				1	813.75
4				2	882.92
5				3	957.97

Cell E2 contains
$A$2*(1+$B$2/$C$2)^$C$2*D2.

The entries in column E have been rounded to two decimal places.

Use a spreadsheet to find the amount in each account over the first 3 years.

1. $P = \$1250$
$r = 7.5\%$
$n = 1$

2. $P = \$1250$
$r = 7.5\%$
$n = 2$

3. $P = \$1250$
$r = 7.5\%$
$n = 3$

4. $P = \$1250$
$r = 7.5\%$
$n = 4$

_____ _____ _____ _____

5. $P = \$1250$
$r = 5.5\%$
$n = 1$

6. $P = \$1250$
$r = 6.5\%$
$n = 1$

7. $P = \$1250$
$r = 7.5\%$
$n = 1$

8. $P = \$1250$
$r = 8.5\%$
$n = 1$

_____ _____ _____ _____

9. Write a brief conclusion that follows from Exercises 1–4.

_____

10. Write a brief conclusion that follows from Exercises 5–8.

_____

Technology
2.3 Building Functions

You can build functions of your own by chaining together a set of operations. For example, the following set of steps will define a function that you can evaluate with a spreadsheet.

Step 1 Begin with a real number x.
Step 2 Multiply x by itself.
Step 3 To the result of step 2 add 4 times the number in step 1.
Step 4 From the result of step 3 subtract 5. Call the result y.

	A	B	C	D
1	STEP 1	STEP 2	STEP 3	STEP 4
2	−2	4	−4	−9
3	−1	1	−3	−8
4	0	0	0	−5
5	1	1	5	0
6	2	4	12	7

Column A contains the values of x.
Cell B2 contains A2*A2.
Cell C2 contains B2+4*A2.
Cell D2 contains C2−5.

Make a spreadsheet for the function defined by each set of steps. Use a reasonable domain for the function.

1. Step 1 Begin with a real number x.
   Step 2 Multiply x by itself.
   Step 3 From the result of step 2 subtract 3 times the number in step 1.
   Step 4 From the result of step 3 subtract 1. Call the result y.

2. Step 1 Begin with a real number x.
   Step 2 Multiply x by 3 times itself.
   Step 3 From the result of step 2 subtract 4 times the number in step 1.
   Step 4 From the result of step 3 subtract 4. Call the result y.

3. Step 1 Begin with a real number x.
   Step 2 Multiply x by twice itself.
   Step 3 To the result of step 2 add 3 times the number in step 1.
   Step 4 From the result of step 3 subtract 2. Call the result y.

4. Step 1 Begin with a real number x.
   Step 2 Multiply x by itself.
   Step 3 To the result of step 2 add the number in step 1.
   Step 4 From the result of step 3 subtract 1. Call the result y.

Plot the ordered pairs from each spreadsheet.

5. Exercise 1   6. Exercise 2   7. Exercise 3   8. Exercise 4

9. Use a spreadsheet to find the minimum value of y from Exercise 1.

_____

Technology
2.4 Domains and Ranges

Sometimes you can easily determine the domain and range of a function. At other times, you may need to use a graphics calculator to get a picture of the function before determining the domain and range. Suppose that you are given the function

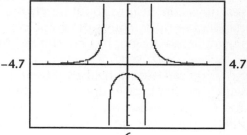

$$f(x) = \frac{1}{(x-1)(x+1)}$$

To determine the domain and range, graph the function on a graphics calculator as shown. From the graph, it appears that the function is defined for all real numbers except -1 and 1. Also from the graph, it appears that $f(x)$ can be any real number greater than zero and less than or equal to -1. So, the range is all real numbers except $-1 < f(x) \leq 0$.

Use a graphics calculator to find the domain and range of each function.

1. $f(x) = x$ **2.** $f(x) = x^2$ **3.** $f(x) = x^3$ **4.** $f(x) = x^4$

_____   _____   _____   _____

5. $f(x) = x^5$ **6.** $f(x) = x^6$ **7.** $f(x) = x^7$ **8.** $f(x) = x^8$

_____   _____   _____   _____

9. $f(x) = \dfrac{1}{(x-1)(x)(x+1)}$ **10.** $f(x) = \dfrac{1}{(x-2)(x-1)(x)(x+1)(x+2)}$

_____   _____

11. Write a brief description of the domains and ranges of the functions in Exercises 1–8. Describe how the graphs are similar and how they are different.

_____

12. Write a brief description of the domains and ranges of the functions in Exercises 9 and 10. Describe how the graphs are similar and how they are different.

_____

Technology
2.5 Slopes and Curves

A graphics calculator and a spreadsheet used together can help you draw conclusions about functions. The graphing calculator display shows the graph of $f(x) = 0.5x^2$. The display suggests that the graph curves up and to the right.

Now you can use a spreadsheet to find the slopes of the line segments that have endpoints $(0, f(0))$ and $(1, f(1))$, $(1, f((1))$ and $(2, f(2))$, and so on. Column C indicates that the slopes of the line segments are positive and increasing.

	A	B	C
1	X	F(X)	SLOPE
2	0	0.0	
3	1	0.5	0.5
4	2	2.0	1.5
5	3	4.5	2.5
6	4	8.0	3.5

Cell C3 contains (B3−B2)/(A3−A2).

Make a spreadsheet like the one shown for each function.
Use $x = 0, 1, 2, 3,$ and 4.

1. $f(x) = 0.25x^2$

_____

2. $f(x) = 0.2x^2$

_____

3. $f(x) = 0.15x^2$

_____

4. $f(x) = 0.1x^2$

_____

5. $f(x) = -x^2$

_____

6. $f(x) = -0.5x^2$

_____

7. $f(x) = -0.25x^2$

_____

8. $f(x) = -0.1x^2$

_____

9. Draw a conclusion based on your spreadsheets from Exercises 1–4.

_____

10. Draw a conclusion based on your spreadsheets from Exercises 5–8.

_____

Technology
2.6 Analyzing Products of Functions

With a graphics calculator, you can see how products of given functions behave. In many cases, products belong to the same family of functions. Let $f(x) = 3x$ and $g(x) = 2x - 3$. Their product $f \cdot g$ is given by the equation:

$$(f \cdot g)(x) = (3x)(2x - 3) = 6x^2 - 9x$$

The graph of the product is shown. Notice that although the graphs of f and g are straight lines, the graph of $f \cdot g$ is a U-shaped curve opening upward.

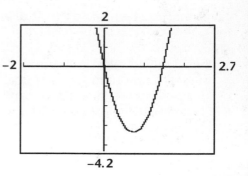

You can also observe that the domain of the product is the set of all real numbers as is the domain of f and g. However, the range of the product is not the set of all real numbers.

Graph the product of each pair of functions. State the domain and range of each product.

1. $f(x) = -2x$ and $g(x) = 4x + 1$

2. $f(x) = 2x + 1$ and $g(x) = 2x - 1$

3. $f(x) = 1.5x + 3$ and $g(x) = 1.5x + 3$

4. $f(x) = x + 2.5$ and $g(x) = 3x - 1.6$

Refer to the graphs drawn in Exercises 1–4. Briefly describe the shape of the graph of $f \cdot g$.

5. $f(x) = ax + b$; $g(x) = cx + d$, where a and c have the same sign.

6. $f(x) = ax + b$; $g(x) = cx + d$, where a and c have opposite signs.

Find the y-intercept of the graph of $f \cdot g$.

7. from Exercise 1

8. from Exercise 2

9. from Exercise 3

10. from Exercise 4

11. If $f(x) = ax + b$ and $g(x) = cx + d$, find the y-intercept of $f \cdot g$. _____

Lesson Activity
2.1 Mathematical Magic

Magicians have performed this "Calendar Magic" trick on stage and in clubs. Now you can too!

1. Find a calendar. Ask a classmate to block off any square of numbers, four numbers on a side. You, the magician, write the "magic" number on a slip of paper and hand it to your classmate. Then turn your back. (In this case, the "magic" number is 60.)

APRIL

S	M	T	W	T	F	S
						1
2	3	4	5	6	7	8
9	10	11	12	13	14	15
16	17	18	19	20	21	22
23	24	25	26	27	28	29
30						

2. Ask your classmate to circle one of the dates in the blocked-off square. Then cross out all the other dates in the same row and same column as the circled date.

3. Ask your classmate to circle another date in the square and again cross out the dates in the same row and same column as this circled date.

4. Have your classmate repeat the process again. Now only one date is left. Have your classmate circle this date and find the sum of all four circled dates.

Here, the sum of the dates is 6 + 12 + 17 + 25 = 60.

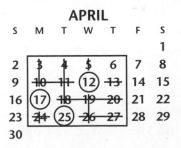

5. Verify that the sum of the circled dates is 76, if the calendar is blocked-off as shown and the same process is performed.

MAY

S	M	T	W	T	F	S
	1	2	3	4	5	6
7	8	9	10	11	12	13
14	15	16	17	18	19	20
21	22	23	24	25	26	27
28	29	30	31			

_____

6. How can you predict the sum of the circled dates?

_____

Lesson Activity
2.2 Patterns With Exponents

If 4^{50} is expanded, what is the digit in the units place?

Look at smaller exponents to find a pattern.

$4^1 = 4$, $4^2 = 16$, $4^3 = 64$, $4^4 = 256$, $4^5 = 1024$, $4^6 = 4096$

1. What are the digits in the units place? What pattern do you notice between odd and even powers of 4 and the units digit?

_____

2. What is the digit in the units place if 4^{50} is expanded? _____

Complete the table.

	x^1	x^2	x^3	x^4	x^5	x^6	Units Digits
1	1	1	1	1	1	1	1
2	2	4	8	16	32	64	2, 4, 8, 6
3	3	9	27				
4							
5							
6							
7							
8							
9							
10							

3. What is the longest cycle for the units digit? _____

4. What is the digit in the units place for 9^{100}? _____

5. How many zeros are there in 10^{50}? _____

6. Which is larger 5^6 or 6^5? Explain. _____

Lesson Activity
2.3 Distance and Time

Karen is traveling by car at a rate of 30 mi/h. For the next minute she accelerates steadily so that 1 minute later she is moving at a rate of 50 mi/h. To determine how far Karen traveled in that minute, you first need to find her average speed.

Her average speed for that minute is

$\frac{30 + 50}{2}$ = 40 mi/h. Therefore, the distance

traveled is $40 \cdot \frac{1}{60} = \frac{2}{3}$ mi/min.

Acceleration is often expressed in feet per second squared (ft/s²). The table shown describes the total distance an object travels going from a state of rest to a steady acceleration of 10 ft/s. Use the table to answer the questions.

Time (s)	Speed at beginning (ft/s)	Speed at end (ft/s)	Average Speed (ft/s)	Total distance (ft)
1	0	10	5	5
2	10	20	15	20
3	20	30	25	45
4	30	40	35	80
5	40	50	45	125
6	50	60	55	180

1. Find the average number of feet traveled during the third second. _____

2. What is the total distance traveled from a state of rest in 3 seconds? _____

3. Use your graphics calculator to graph the ordered pairs (t, d).

4. Find a formula that describes the graph. _____

5. Is this formula a function? Explain. _____

6. Give the domain and range. _____

7. Find the distance, in feet, traveled in 1 minute. _____

Lesson Activity
2.4 Energy and Time

The energy value of food is measured in units of heat energy, or calories. A food calorie is the amount of heat required to raise the temperature of 1 kilogram of water 1 degree Celsius.

Ed notices that during a football season he loses weight, but in the winter he gains weight. He wants to maintain a constant level of body weight. The table shown describes the calories used per minute for various activities.

Calories used per minute	Activity	Calories used per minute	Activity
10.1	Football	2.8	Badminton
8.6	Basketball	8.0	Rowing
4.8	Ping pong	2.6	Sailing
12.1	Swimming	3.0	Playing pool
5.5	Golf	4.0	Dancing
7.0	Tennis	3.0	Horseback riding
8.1	Bowling	8.0	Cycling

1. How many calories will Ed use by playing football 4 hours per day? _____

2. If Ed rides a bicycle for 1.5 hours per day for 5 days, how many calories

    will he use? _____

3. Choose one activity and express the total number of calories used per hour for that activity as a function of time using function notation.

    _____

4. Use your graphics calculator to graph the function.

5. What is the domain of the function? _____

6. What is the range of the function? _____

7. What conclusion can you draw about the relationship between calories

    and time? _____

8. Resting in bed uses 1.2 calories per minute. Find the number of calories
    used in 8 hours of sleep.   _____

Lesson Activity
2.5 Fluid Flow and Time

The ancient Egyptians devised water clocks called clepsydras for telling time, especially at night, when sun dials were impractical. In a clepsydra, water dropped out through a hole in the base of a container. As the water level fell, so did a float on the surface. The float was attached to a pointer that marked the passing of the hours on a scale.

Place water in a plastic container. Make a small hole near the base of the container. Let water drip into a measuring cup placed below the container. Measure the amount of water collected at 5-minute intervals. Mark the water level on the container as well.

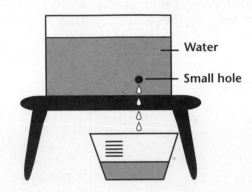

1. Complete the table.

Time passed (min)	5	10	15	20	25	30
Amount of water (milliliters per min)						

2. Write a linear function relating the time passed and the amount of water that drips out of the container.

_____

3. Estimate where the water level in the container will be after 1 hour. _____

4. Find the constant rate of change. _____

5. Is the function a direct variation? Explain.

_____

6. What is the slope of the function? _____

7. What is the y-intercept? _____

8. Is the function an increasing or decreasing function? _____

Lesson Activity
2.6 Demand Curve and Time

Mike sells compact discs. He wants to determine how many compact discs customers will purchase in a month at various possible prices. He records the number of discs purchased by Bruno and Lauren over a one-month period at varying prices. He also records the same information for all other customer purchases over the same one-month period.

The table shows the data collected by Mike.

Price of CDs (dollars)	3	4	5	6	7	8	9	10	11	12	13	14	15	16
Y_1 Bruno	10	9	8	7	6	5	4	3	2	1	0	0	0	0
Y_2 Lauren	8	7	6	5	4	3	2	1	0	0	0	0	0	0
Y_3 All other customers	182	168	154	140	126	112	98	84	70	56	42	28	14	0

1. Use your graphics calculator to enter the data points. Make a scatter plot and find the equation for the line of best fit for each of the three functions.

   _____

2. How many $9 CDs did Bruno buy that month? _____

3. As the price of CDs increased, did Lauren buy more or fewer discs

   during the month? _____

4. What represents the dependent variable for each of the three functions? _____

5. In economics, the Law of Demand assumes that as the price rises, the quantity demanded falls; as price falls, the quantity demanded rises. Do the functions Y_1, Y_2, and Y_3 support the assumption? Explain.

   _____

6. The total market demand curve Y_4 is the result of adding, at each value of the independent variable, the values of the three dependent variables Y_1, Y_2, and Y_3. Find the equation for Y_4.

   _____

7. What is the relationship between the domains and ranges of Y_1, Y_2,

   and Y_3, and the domain and range of Y_4? _____

Assessing Prior Knowledge
2.1 Exploring Numbers

Evaluate each expression if $x = 0.2$, $y = -3$, and $z = \frac{1}{3}$.

1. $x - z$ _____
2. $y + z$ _____
3. $\frac{1}{x + y}$ _____
4. yz _____

- -

Quiz
2.1 Exploring Numbers

Solve.

1. $\frac{27}{2} \cdot \frac{6}{9}$ _____

2. $\frac{5}{6} + \frac{4}{9}$ _____

3. $\frac{2}{3} - \frac{1}{2}$ _____

4. $\frac{3}{4} \div 4$ _____

5. $\frac{4x - 4}{2x + 2} \cdot \frac{2x + 2}{2x - 2}$ _____

6. $\frac{x + 3}{3x} \cdot \frac{9}{3x + 9}$ _____

7. $\frac{x}{9y + 6} \div \frac{2x(3y + 2)}{(3y + 2)^2}$ _____

8. $\frac{x^2}{x + 2} \div \frac{3x}{3x + 6}$ _____

9. $\frac{3x + 4}{x} + \frac{5x}{x^2}$ _____

10. $\frac{-1(4 - x)}{3} - \frac{1x}{2}$ _____

11. $\frac{12y}{3(y - 1)} \div y^2$ _____

12. $\frac{5x - 30}{3x} \div (2x - 12)$ _____

Assessing Prior Knowledge
2.2 Exploring Properties of Exponents

Simplify.

1. 3^3 _____

2. 7^2 _____

3. 2^5 _____

4. 4^3 _____

- -

Quiz
2.2 Exploring Properties of Exponents

Simplify.

1. $(4^8)(4^{-6})$ _____

2. $(4^2)(4^{-3})$ _____

2. $(3^9)^{\frac{1}{3}}$ _____

4. $\dfrac{(x^2 - x)^0}{2}$ _____

5. $(y^8)^{\frac{1}{4}}$ _____

6. $\left(\dfrac{4x^2}{-5y^3}\right)^{-3}$ _____

7. $\dfrac{-3x^2y^{-3}z}{4x^3yz^{-2}}$ _____

8. $(6a^{-5}b^2)(2a^{-5}b^3)^{-1}$ _____

9. $(3x^{-1}y^3)(5x^4y^{-2})$ _____

10. $(-2mn^2)^{-3}(2m^2n^{-4})$ _____

11. $(3^6)^{0.5}$ _____

12. $3(6x^3y^{-1})^2$ _____

Assessing Prior Knowledge
2.3 Definition of a Function

1. Complete the table for $y = 2x^2 + 1$.

x	-3	-2	-1	0	1	2	3
y							

2. Are there any *x*-values that correspond to the same *y*-values? If so, name them.

_____

- -

Quiz
2.3 Definition of a Function

Determine whether or not each set of ordered pairs represents a function. If not, explain.

1. {(1, 2), (3, 4), (1, 6), (5, 8)}

_____

2. {(−2, 2), (−1, 1), (1, −1), (2, −2)}

_____

Determine whether or not each table represents a function. If not, explain.

3.

x	5	6	7	8
y	1	2	3	4

_____

4.

x	3	6	5	2
y	1	2	1	2

_____

Determine whether or not each graphed relation is a function. If not, explain.

5.

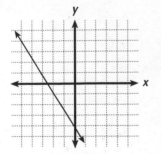

6.

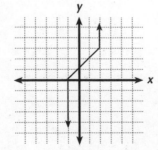

7.

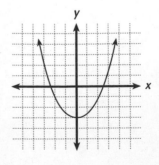

_____

HRW Advanced Algebra

NAME _____ CLASS _____ DATE _____

 # Mid-Chapter Assessment
Chapter 2 (Lessons 2.1–2.3)

Write the letter that best answers the question or completes the statement.

_____ **1.** When simplified, $\frac{4}{3} + \frac{2}{9} =$

         a. $\frac{1}{2}$ **b.** $\frac{14}{9}$ **c.** $\frac{2}{9}$ **d.** $\frac{2}{3}$

_____ **2.** Which set of ordered pairs represents a function?

         a. $\{(4, -4), (3, -3), (2, -2), (1, -1)\}$ **b.** $\{(-4, 1), (-2, 2), (-2, 1), (4, 2)\}$
         c. $\{(-4, 1), (-2, 2), (-4, -1), (2, 2)\}$ **d.** $\{(1, 2), (2, 2), (1, 3), (4, 4)\}$

_____ **3.** Which of the following tables does not represent a function?

         a.

x	1	2	3	4
y	4	3	5	6

         b.

x	-2	-1	1	2
y	6	3	2	7

         c.

x	5	1	6	9
y	3	3	3	3

         d.

x	4	4	4	4
y	-4	-2	2	4

_____ **4.** What is the range of the function $y = x^2 - 1$ when the domain is $\{2, 3, 4, 5\}$?

         a. $x = \{3, 8, 15, 24\}$ **b.** $3 \le x \le 24$
         c. $y = \{3, 8, 15, 24\}$ **c.** $3 \le y < 24$

Simplify.

5. $\frac{6}{5} - \frac{1}{7}$ _____   **6.** $\frac{5}{9} \cdot \frac{-3}{5}$ _____

7. $\frac{4x - 4}{8} + \frac{x}{5}$ _____   **8.** $\frac{9^{-2}}{18^{-1}}$ _____

9. $\frac{x^2}{3x^2 + 6x} \div \frac{2x - 4}{x^2 - 4}$ _____   **10.** $(4^{-3})(4^5)$ _____

11. $4x^{-3}y^2z^4(3x^2y^4z^{-2})^{-2}$ _____   **12.** $\frac{(5a^2b^3c^{-1})^0}{(-2a^3bc^2)^{-1}}$ _____

NAME _____ CLASS _____ DATE _____

Assessing Prior Knowledge
2.4 Using Notation to Represent Functions

1. Given $y = x - 3$, find y if $x = 6$. _____

2. Given $y = 4x + 5$, find y if $x = -2$. _____

- -

NAME _____ CLASS _____ DATE _____

Quiz
2.4 Using Notation to Represent Functions

If $f(x) = x^2 - 4$, find each value.

1. $f(-2)$ _____   **2.** $f(a + 2)$ _____

Let $g(x) = x^2 - 6x + 3$. Find each value.

3. $g(3)$ _____   **4.** $g(a - b)$ _____

5. If $f(x) = x^2 - 1$, graph $f(x)$ on the grid provided.

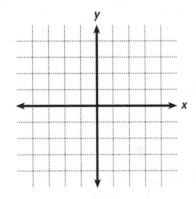

6. What is the domain of $f(x)$? _____

7. What is the range of $f(x)$? _____

8. What is the vertex of $f(x)$? _____

9. Identify the independent and dependent variables.

_____

Assessing Prior Knowledge
2.5 Linear Functions

Let $f(x) = -\frac{3}{4}x + 1$.

1. Find $f(4)$. _____

2. Find $f(-4)$. _____

3. Find $f(2)$. _____

4. Find $f(0.5)$. _____

- -

Quiz
2.5 Linear Functions

Write the linear function f using the given information. Then tell whether f is increasing, decreasing, or neither.

1. $m = \frac{1}{2}; b = 2$

2. $m = -1; f(0) = -1$

_____

_____

3. $(-1, -2)$ and $(1, 2)$

4. $(4, 3)$ and $(5, 2)$

_____

_____

5.

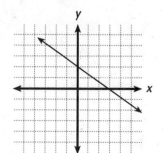

_____

6.

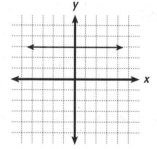

_____

Assessing Prior Knowledge
2.6 Exploring Operations With Functions

Simplify.

1. $(x - 2) + (x + 5)$ _____

2. $(3x + 4) + (-2x - 6)$ _____

3. $(x + 2)(x - 3)$ _____

4. $(2x - 1)(x + 5)$ _____

- -

Quiz
2.6 Exploring Operations With Functions

Let $f(x) = 1 - x^2$ and $g(x) = 3x - 1$. Find the domain and range for each of the following.

1. g _____

2. f _____

3. $g + f$ _____

4. $f \cdot \dfrac{1}{g}$ _____

5. $g \cdot \dfrac{1}{f}$ _____

6. $g \cdot f$ _____

Use the given functions to find the combinations $(f \div g)(x)$ and $(g \cdot f)(x)$.

7. $f(x) = 2x^2 - 2x$

$g(x) = 4x^3$

_____

_____

8. $f(x) = 2x^3 - 6x^2$

$g(x) = \dfrac{1}{x - 3}$

_____

_____

Use the given functions to find the combinations $(f - g)(x)$ and $\left(g \cdot \dfrac{1}{f}\right)(x)$.

9. $f(x) = 4x^3$
$g(x) = 2x - 8$

_____

_____

10. $f(x) = 9x - 3$
$g(x) = 3x^2 - 12x - 6$

_____

_____

Chapter Assessment
Chapter 2, Form A, page 1

Write the letter that best answers the question or completes the statement.

_____ 1. Which of the following represents a linear function such that $m = \frac{-1}{2}$ and $f(2) = 5$?

    a. $f(x) = -2x + 5$ **b.** $f(x) = -2x + 6$

    c. $f(x) = 2x - \frac{1}{2}$ **d.** $f(x) = -\frac{1}{2}x + 6$

_____ 2. The linear function containing the points (5, 2) and (3, 4) is

    a. $f(x) = 5x - 2$ **b.** $f(x) = -x + 7$

    c. $f(x) = 5x + 7$ **d.** $f(x) = -x - 7$

_____ 3. Each of the graphs below represents a function except

    a.

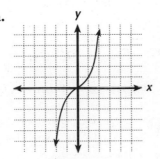

    b.

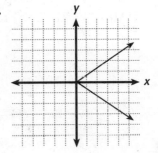

    c.

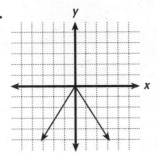

    d.

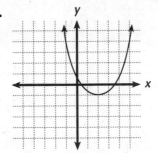

_____ 4. $\frac{5}{4} + \frac{2}{3} =$

    a. $\frac{23}{12}$ **b.** $\frac{7}{12}$ **c.** 10.7   **d.** 1

_____ 5. When written in simplest form, $[(x^2)^9]^{\frac{1}{6}} =$

    a. x^5 **b.** x^3 **c.** x^{12} **d.** $\frac{x^{11}}{6}$

_____ 6. $\frac{2x - 2}{(4x^2 - 4x)^0} =$

    a. $\frac{1}{2x}$ **b.** $\frac{x - 2}{x(2x - 2)}$ **c.** $2x - 2$ **d.** undefined

Chapter Assessment
Chapter 2, Form A, page 2

_____ 7. $\left(\dfrac{-2x^3}{3x^4}\right)^{-2} =$

 a. $\dfrac{4}{9x^{-2}}$ **b.** $\dfrac{-4}{9x^{-2}}$ **c.** $\dfrac{-9x^2}{4}$ **d.** $\dfrac{9x^2}{4}$

_____ 8. Which set of ordered pairs does not represent a function?

 a. $\{(5, 6), (4, 5), (5, -6), (3, 4)\}$ **b.** $\{(-2, 1), (-1, 0), (0, 1), (1, 2)\}$
 c. $\{(1, 4), (2, 3), (3, 2), (4, 1)\}$ **d.** $\{(4, 4), (3, 4), (2, 4), (1, 4)\}$

For Exercises 9–12 let $f(x) = -2x + x^2$ and $g(x) = x - 9$.

_____ 9. Which of the following is the domain of $\dfrac{1}{g}$?

 a. all real numbers **b.** $x > 9$
 c. all real numbers except $x \neq -9$ **d.** all real numbers except $x \neq 9$

_____ 10. Which of the following is the vertex of $f(x)$?

 a. $(1, 1)$ **b.** $(0, -1)$ **c.** $(1, 0)$ **d.** $(1, -1)$

_____ 11. Which of the following is the value of $g(a + b)$?

 a. $-2a - 2b + (a + b)^2$ **b.** $9 + a + b$
 c. $-9a - 9b$ **d.** none of the above

For Exercises 12–14, let $f(x) = 4x^2 - 1$ and $g(x) = \dfrac{2x}{x + 2}$.

_____ 12. Which of the following is the range of f?

 a. all real numbers **b.** $y \geq 0$ **c.** $y \geq -1$ **d.** $y \leq -1$

_____ 13. Which of the following are the domain and range of $g \cdot \dfrac{1}{f}$?

 a. $D =$ all real numbers except $x \neq -2$ **b.** $D =$ all real numbers
 $R =$ all real numbers $R = y \geq 0$

 c. $D =$ all real numbers except $x \neq -2, x \neq \dfrac{1}{2}, x \neq -\dfrac{1}{2}$
 $R = y \geq 0$

 d. $D =$ all real numbers except $x \neq -2, x \neq \dfrac{1}{2}, x \neq -\dfrac{1}{2}$
 $R =$ all real numbers

_____ 14. Which of the following functions is equal to $f \cdot \dfrac{1}{g}$?

 a. $\dfrac{8x^3 - 2x}{x + 2}$ **b.** $\dfrac{4x^3 + 8x^2 - x - 2}{2x}$

 c. $\dfrac{2x}{4x^3 + 8x^2 - x - 2}$ **d.** none of the above

Chapter Assessment
Chapter 2, Form B, page 1

Determine the linear function containing the given points.

1. (2, 3) and (3, 5) _____

2. (6, 1) and (4, 2) _____

State whether each graphed relation is a function.

3.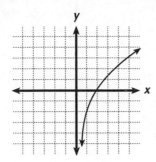

4.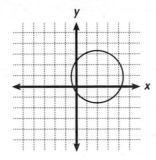

5.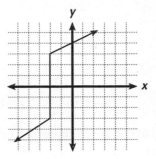

_____ _____ _____

Write the linear function f using the given information.

6. $m = -2; b = -1$

7. $m = \frac{1}{3}; f(0) = 2$

_____ _____

Simplify each expression.

8. $\dfrac{x + 2}{2} \div \dfrac{3(x + 2)}{x^2 - x}$ _____

9. $\dfrac{9x^2 - 16}{3x - 4} \bullet \dfrac{x + 4}{3x + 4}$ _____

10. $\dfrac{x - 1}{x + 1} + \dfrac{1}{x - 1}$ _____

11. $\dfrac{6x^4 y^{-3} z}{5x^2 y^{-1} z^{-2}}$ _____

State whether the set of ordered pairs represents a function.

12. $\{(-3, 3), (-1, 1), (1, -1), (3, -3)\}$ _____

13. $\{(2, 6), (3, 5), (2, 4), (4, 3)\}$ _____

For Exercises 14–18, let $f(x) = x^2 - 4$.

14. Graph $f(x)$ on the grid provided.

15. What is the domain of $f(x)$?

_____

16. What is the range of $f(x)$? _____

17. Find the value of $f(b + 1)$. _____

18. What is the vertex of $f(x)$? _____

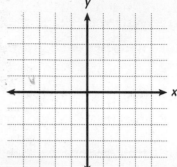

State whether each table represents a function.

19.

x	6	5	4	6
y	1	3	5	7

20.

x	2	3	4	5
y	4	4	4	4

_____   _____

For Exercises 21–22, let $g(x) = \dfrac{x}{x^3 - 27}$ and $f(x) = 3x^2$.

21. Find $g \cdot \dfrac{1}{f}$. _____
22. What is the domain of $g \cdot \dfrac{1}{f}$? _____

23. Let $g(x) = 2x + 3$ and $f(x) = -5x - 4$. Graph $f + g$ on the grid provided.

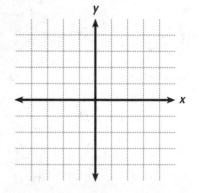

24. Let $g(x) = 6x^2$ and $f(x) = -3$. Graph $g \cdot \dfrac{1}{f}$ on the grid provided.

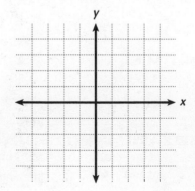

Alternative Assessment
Exploring Rules of Exponents, Chapter 2, Form A

TASK: To identify the rules of exponents, and use these rules to perform operations with exponents

HOW YOU WILL BE SCORED: As you work through the task, your teacher will be looking for the following:

- how well you understand the definition of an exponent
- whether you can perform operations and simplify expressions using the rules of exponents
- how effectively you can communicate your responses in writing

1. Create two exponents m and n. Describe the types of numbers you can choose to represent the base a, and the exponents m and n.

   _____

2. Use your two exponents from Exercise 1. Does $a^m \cdot a^n = a^{m \cdot n}$?
   Does $\frac{a^n}{a^m} = a^{n-m}$? Explain why or why not. If the statement is false, explain how to change the false statement into a true statement.

   _____

If possible, simplify. Then write the general rule that applies to each statement.

3. $(3xy)^{-1}$ _____

4. $(p^2 \cdot p^{-1})^0$ _____

Choose any values for a. Substitute these values in the following equations and evaluate.

5. $(1 + a)^{-1} = -1 + a^{-1}$ _____

6. $1 + a^{-1} = \frac{1}{1 + a}$ _____

7. $1^{-1} + a^{-1} = 1 + \frac{1}{a}$ _____

8. $(1 + a)^{-1} = \frac{1}{1 + a}$ _____

9. Which equations are true?

   _____

10. Write a general rule that would apply to $(1 + a)^{-1}$, where a represents any real number.

    _____

SELF-ASSESSMENT: Explain how you could use your graphics calculator to find the general rule that would apply to $(1 + a)^{-1}$, where a represents any real number.

Alternative Assessment
Exploring Operations with Functions, Chapter 2, Form B

TASK: To describe the operations of addition, subtraction, multiplication, and division of functions

HOW YOU WILL BE SCORED: As you work through the task, your teacher will be looking for the following:

- how well you use properties of functions to add, subtract, multiply, and divide functions
- how well you can perform the indicated operation

Let $f(x) = x + 2$ and $g(x) = x^2 + 3x + 2$.

1. Describe how the domain and range of the sum function, $(f + g)(x)$, compares with the domain and range of the individual functions, f and g. Then find $(f + g)(x)$.

   _____

   _____

2. Find $(f - g)(x)$. Is $(f - g)(x)$ always equal to $f(x) - g(x)$? Explain.

   _____

   _____

3. Explain how the domain and range of $(f \cdot g)(x)$ are constructed from f and g. Then find $(f \cdot g)(x)$. Is $(f \cdot g)(x) = (g \cdot f)(x)$? Use an example to explain your result.

   _____

   _____

4. Explain what special domain and range restrictions occur when the quotient $\frac{f(x)}{g(x)}$ is found. Then find the quotient.

   _____

   _____

SELF-ASSESSMENT: Use your graphics calculator to explain what happens when two functions are combined by the operations of addition, subtraction, multiplication, and division.

Practice & Apply
3.1 Exploring Symmetry

Determine whether *P* and *Q* are symmetric with respect to the *x*-axis or the *y*-axis.

1. $P(1, 3); Q(1, -3)$

2. $P(-1, 4); Q(1, 4)$

3. $P(3, -3); (-3, -3)$

_____   _____   _____

For each point, find the coordinates of the corresponding point symmetric with respect to the *x*-axis and the *y*-axis.

4. $A(-1, 2)$

5. $B(0, -2)$

6. $C(-2, 0)$

_____   _____   _____

Graph a triangle with vertices *A*(2, 1), *B*(3, 2), and *C*(1, 3).

7. Determine the coordinates of the image points that are symmetric to your pre-image points with respect to the

    y-axis. _____

8. Plot the image points on the same coordinate plane as the pre-image triangle. Draw the image triangle. What does

    your new triangle resemble? _____

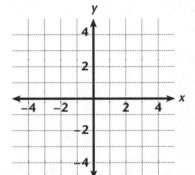

For each point, find the coordinates of the corresponding point symmetric with respect to the line *y* = *x*.

9. $A(4, -5)$ _____

10. $B(0, -2)$ _____

11. $C(8, 3)$ _____

12. Describe how you can determine if two points are symmetric with

    respect to the line $y = x$. _____

Graph a square with vertices *A*(3, 0), *B*(5, 0), *C*(5, 2) and *D*(3, 2).

13. Determine the coordinates of the image points that are symmetric to your pre-image points with respect to the line

    $y = x$. _____

14. Plot the image points on the same coordinate plane as the pre-image square. Draw the image square. What does your

    new square resemble? _____

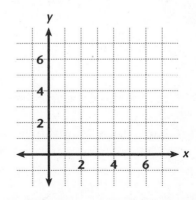

NAME _____ CLASS _____ DATE _____

Practice & Apply
3.2 Inverse Functions

Find the inverse of each function. Identify which inverses are functions.

1. $\{(4, 3), (2, 2), (0, 1), (-2, 0)\}$ _____

2. $\{(-3, -5), (-2, 1), (1, 3), (2, 6)\}$ _____

3. $\{(6, -2), (5, -1), (3, 0), (5, -2)\}$ _____

4. $\{(6, 8), (4, 6), (2, 3), (1, 5)\}$ _____

Use your graphics calculator to determine if f and g are inverse functions.

5. $f(x) = \dfrac{2x + 1}{3}$ and $g(x) = \dfrac{3x - 1}{2}$ _____

6. $f(x) = 5x - 1$ and $g(x) = \dfrac{x + 1}{5}$ _____

7. $f(x) = x + 1$ and $g(x) = \dfrac{1}{x + 1}$ _____

8. $f(x) = \dfrac{x}{3} - 2$ and $g(x) = 3x + 2$ _____

Find the inverse of each function.

9. $f(x) = 7x + 4$ _____

10. $f(x) = \dfrac{x}{5} - 6$ _____

11. $f(x) = \dfrac{3x - 1}{2}$ _____

In Exercises 12–15, $f(x) = 3x + 4$.

12. What are the slope and y-intercept of f? _____

13. Find the inverse of $f(x) = 3x + 4$. _____

14. What are the slope and the y-intercept of f^{-1}? _____

15. What is the relationship between the slope of f and the slope of f^{-1}?

_____

_____

HRW Advanced Algebra

NAME _____ CLASS _____ DATE _____

Practice & Apply
3.3 Composition of Functions

For Exercises 1–4, let $f(x) = x - 1$ and $g(x) = 3x + 2$.

1. Find the domain and range of f and g. _____

2. Find $f \circ g$ and $g \circ f$. _____

3. Is composition a commutative operation? _____

4. What are the domain and range of $f \circ g$ and $g \circ f$? _____

For Exercises 5–12, let $f(x) = 2x + 3$ and $g(x) = x - \frac{1}{2}$. Find each value.

5. $f \circ g$ _____

6. $g \circ f$ _____

7. $f \circ f$ _____

8. $g \circ g$ _____

9. $(f \circ g)(-3)$ _____

10. $(g \circ g)(2)$ _____

11. $(g \circ f)(0)$ _____

12. $(f \circ f)(1)$ _____

13. The circumference C of a circle and the diameter d are given by
$C = f(d) = \pi d$ and $d = g(r) = 2r$. Find $(f \circ g)(r)$. Interpret the result.

_____

For Exercises 14–22, let $f(x) = x^2 - 2$, $g(x) = -3x + 1$, and $h(x) = 2x$. Find each value.

14. $f \circ g$ _____

15. $g \circ f$ _____

16. $f \circ h$ _____

17. $h \circ g$ _____

18. $f \circ (g \circ h)$ _____

19. $(f \circ h)(2)$ _____

20. $(h \circ g)(0)$ _____

21. $(g \circ f)(-1)$ _____

Write $h(x)$ as the composition of two functions f and g for which $(f \circ g)(x) = h(x)$.

22. $h(x) = (x + 5)^2$ _____

23. $h(x) = 3x - 4$ _____

Let $f(x) = x - 1$ and $g(x) = 2x + 4$. Jim found $f(g(x))$ as $2x + 3$. Jim then evaluated $f(g(x))$ for $x = 2$. Megan evaluated $g(x)$ for $x = 2$ and then used the result to evaluate $f(x)$.

24. Find $f(g(2))$ using Jim's method. _____

25. Find $f(g(2))$ using Megan's method. _____

26. Explain the result. _____

Practice & Apply
3.4 The Absolute Value Function

Simplify.

1. $|-1.5| - |2|$ _____

2. $|3.75| - |3|$ _____

3. $|-4| + |-4|$ _____

4. $-|-6| + |-8|$ _____

5. $|5.64| - |-2.3|$ _____

6. $|-12| - |9.4|$ _____

Solve each equation for x using the definition of absolute value.

7. $|2x - 9| = 7$ _____

8. $|x + 2| = 5$ _____

9. $|8 + 4x| = 8$ _____

10. $|2 - x| = 4$ _____

11. $|4x| = 16$ _____

12. $3|x + 1| = 9$ _____

13. $|7 - 8x| = -15$ _____

14. $\left|1 - \frac{2}{3}x\right| = 5$ _____

15. $\left|\frac{3}{4}x + 3\right| = 9$ _____

16. $|x - 4| = 2x - 4$ _____

17. $|2x + 6| = x + 6$ _____

18. $|2 - 3x| = x + 4$ _____

Use your graphics calculator to graph each function. Give the domain and range.

19. $f(x) = |x| + 2$

20. $f(x) = |3x| - 1$

21. $f(x) = -|1 - x|$

_____   _____   _____

Let $f(x) = |x + 3|$ and $g(x) = |x| - 3$. Use your graphics calculator in these exercises.

22. Graph f and g on the same coordinate plane.

23. Compare the graphs noting similarities and differences.

_____

Given that f is an absolute value function, determine the function represented by each graph.

24.

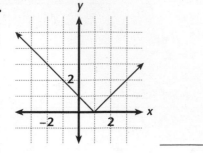

_____

25.

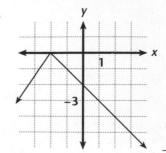

_____

Practice & Apply
3.5 Step Functions

Simplify.

1. [4.35] _____   2. [−2] _____   3. [6] _____   4. [−3.15] _____

5. [−4.35] _____   6. [−3.4] _____   7. [1.2] _____   8. [−7] _____

9. [6.5] + [−2.3] _____   10. [−8] − [−4.1] _____

11. [7.8] − [−5.9] _____   12. [−7] − [−4] _____

13. [8.7] − [6.2] _____   14. [−5] + [5] _____

15. ⌈−2.5⌉ _____   16. ⌈6.15⌉ _____   17. ⌈5.990⌉ _____   18. ⌈−6.01⌉ _____

19. ⌈−14.59⌉ _____   20. ⌈−2.9⌉ _____   21. ⌈$\sqrt{3}$⌉ _____   22. ⌈−10.3⌉ _____

23. ⌈−5.2⌉ − ⌈5.99⌉ _____   24. ⌈7.1⌉ − ⌈−6.5⌉ _____   25. ⌈$\sqrt{2}$⌉ − [4.9] _____

26. ⌈−4.2⌉ − ⌈−6.4⌉ _____   27. ⌈−1.9⌉ + ⌈−2.9⌉ _____   28. [−4.1] − [3.6] _____

Let $f(x) = -2[x]$ and $g(x) = [-2x]$. Use your graphics calculator.

29. Graph f and g on the same coordinate plane.

30. Describe the similarities and differences between the graphs of these

funcions. _____

The cost for parking a car in a parking lot is $2 for the first half hour. For
each additional half hour or part, there is an additional $1 charge.

31. Complete the table.

time (in hours)	$0 < t \le \frac{1}{2}$	$\frac{1}{2} < t \le 1$	$1 < t \le \frac{3}{2}$	$\frac{3}{2} < t \le 2$	$2 < t \le 2\frac{1}{2}$	$2\frac{1}{2} < t \le 3$
cost						

32. Graph the data.

33. Explain why the graph is a function.

_____

34. What type of step function is represented? _____

35. Write a step function for the cost of parking a car.

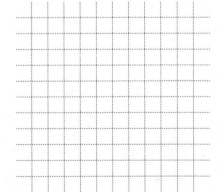

_____

Practice & Apply
3.6 Parametric Equations

Combine each system into one function. Check your result by graphing each parametric system on your graphics calculator.

1. $\begin{cases} x(t) = t - 3 \\ y(t) = t + 6 \end{cases}$
       2. $\begin{cases} x(t) = 3t + 1 \\ y(t) = -2t \end{cases}$
       3. $\begin{cases} x(t) = 1 - t \\ y(t) = t \end{cases}$

_____   _____   _____

4. $\begin{cases} x(t) = t - 5 \\ y(t) = 3t + 2 \end{cases}$
       5. $\begin{cases} x(t) = 2t - 1 \\ y(t) = 5 - 6t \end{cases}$
       6. $\begin{cases} x(t) = 2t + 1 \\ y(t) = 1 - 2t \end{cases}$

_____   _____   _____

7. Write parametric equations for the line through (2, 3) and (−1, 4).

_____

A plane descends at 125 feet per second horizontally and 30 feet per second vertically.

8. Give a parametric representation of the airplane's descent path. _____

9. Use your graphics calculator to determine how far the plane has

traveled in 5 minutes. _____

10. Combine your parametric system into a linear function and check it by graphing.

_____

A baseball is thrown from a height of 6 feet with an initial velocity of 50 feet per second at a 60° angle with the ground. The following system of parametric equations describes the ball's path.

$$\begin{cases} x(t) = 25t \\ y(t) = 6 + 25\sqrt{3}t - 16t^2 \end{cases}$$

11. What is the maximum height attained by the ball? _____

12. How long does it take for the ball to reach its maximum height? _____

13. How far does the ball travel before it hits the ground? _____

14. How long does it take for the ball to hit the ground? _____

Enrichment
3.1 Code of Symmetry

Choose the graph that represents the image of the first graph when it is reflected over the line $y = x$. Write the corresponding letter on the blank provided.

_____ **1.**

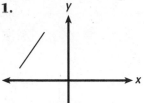   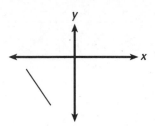   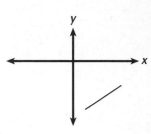

P   I

_____ **2.**

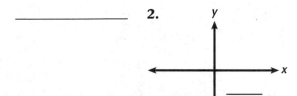

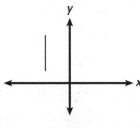   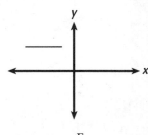

M   E

_____ **3.**

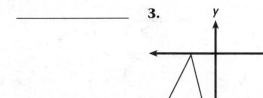

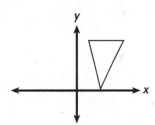   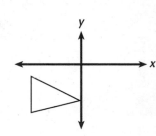

R   A

_____ **4.**

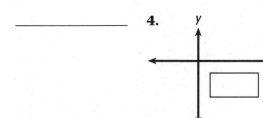

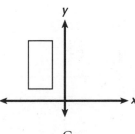   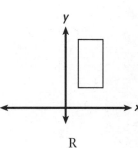

G   R

_____ **5.**

   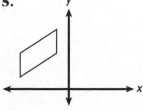

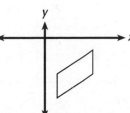   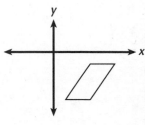

N   E

Enrichment
3.2 Shades of Functions

Shade each box that contains a function and its inverse function. What do you see?

_____

$y = 3x + 7$ $y = 7x + 3$	$y = 2x + 4$ $y = -2x + 4$	$y = -3x + 1$ $y = 3x + 1$	$y = 2x + 1$ $y = \frac{1}{2}x - \frac{1}{2}$	$y = x - 2$ $y = x + 2$	$y = x^3$ $y = \sqrt[3]{x}$	$y = 2x + 1$ $y = \frac{1}{2}x + 1$
$y = x^2$ $y = \sqrt{x}$	$y = 4$ $x = 4$	$y = x^3 + 1$ $y = \sqrt[3]{x - 1}$	$y = \|x\|$ $y = x^2$	$2y = x$ $y = -2x$	$x + 2y = 6$ $2(x + y) = 6$	$y = 3x$ $y = \frac{1}{3}x$
$y = 4x + 1$ $y = -4x + 1$	$y = x^4$ $y = \sqrt[4]{x}$	$y = x^3 - 4$ $y = \sqrt[3]{x + 4}$	$y = 7x + 3$ $y = \frac{1}{7}x + 3$	$y = \frac{2}{3}x + 1$ $y = -\frac{2}{3}x - 1$	$y = \frac{x + 1}{4}$ $y = \frac{x - 1}{4}$	$y = \frac{2x - 3}{5}$ $y = \frac{-2x + 3}{5}$
$y = \frac{2x + 1}{4}$ $y = \frac{4x + 1}{2}$	$y = \frac{4x + 5}{3}$ $y = \frac{4x - 3}{5}$	$y = \frac{1}{2}x - 3$ $y = 2x + 6$	$y = x^3$ $y = \sqrt{x}$	$y = x^2$ $y = \sqrt[3]{x}$	$y = 5x + 3$ $y = \frac{x + 3}{8}$	$x - y = 4$ $x + y = 4$
$y = \frac{4}{5}x + \frac{1}{5}$ $y = \frac{5}{4}x - \frac{1}{4}$	$x + y = 7$ $x + y = 7$	$y = \frac{3}{4}x + 1$ $y = \frac{4}{3}x - \frac{4}{3}$	$3x + 4y = 12$ $4x + 3y = 12$	$2x + 3y = 8$ $3x + 2y = 8$	$2x + y = 3$ $2x - y = 8$	$x + y = 3$ $x + y = \frac{1}{3}$
$3x - 2y = 6$ $3x + 2y = 6$	$2x - 5y = 10$ $5x - 2y = 10$	$7x = 4y$ $4x = 7y$	$\frac{x}{3} - \frac{y}{4} = 1$ $\frac{x}{4} + \frac{y}{3} = 1$	$3x - 4y = 20$ $4x - 3y = 20$	$2x + y = 8$ $2x - y = 8$	$y = \frac{2x - 1}{3}$ $y = \frac{3x - 1}{2}$
$y = \frac{8x - 2}{3}$ $y = \frac{2x - 8}{3}$	$y = \frac{3x + 1}{4}$ $y = \frac{x - 3}{4}$	$y = 2x + 6$ $y = \frac{1}{2}x - 3$	$y = \frac{2x - 1}{5}$ $y = \frac{x - 2}{5}$	$y = \frac{x^3 - 1}{4}$ $y = \frac{3x - 1}{4}$	$2x - 7y = 9$ $y = \frac{2x + 9}{7}$	$y = \frac{3x - 5}{19}$ $y = \frac{5x + 3}{19}$
$y = \frac{4x - 5}{7}$ $y = \frac{7x - 5}{4}$	$y = \frac{9}{4}x$ $y = -\frac{4}{9}x$	$y = -2x + 6$ $y = -\frac{1}{2}x + 3$	$y = \frac{x + 1}{4}$ $y = \frac{4x - 1}{2}$	$y = \frac{3x - 1}{5}$ $y = \frac{3x - 4}{6}$	$y = \frac{7x + 2}{5}$ $y = \frac{3x + 1}{6}$	$y = \frac{4x - 7}{8}$ $y = \frac{5x + 3}{2}$

Enrichment
3.3 Commuting

Use composition of functions to determine whether each pair of functions is commutative. Write the letter corresponding to your answer on the blank provided.

_____ 1. $f(x) = 2x - 1$ $g(x) = \dfrac{x + 1}{2}$ Yes: T   No: I

_____ 2. $f(x) = x - \dfrac{3}{4}$ $g(x) = 4x + 3$ Yes: H   No: N

_____ 3. $f(x) = -3x + 6$ $g(x) = \dfrac{x + 6}{-3}$ Yes: V   No: E

_____ 4. $f(x) = x + \dfrac{4}{7}$ $g(x) = 7x - 4$ Yes: T   No: E

_____ 5. $f(x) = \dfrac{2x - 1}{4}$ $g(x) = \dfrac{4x - 1}{2}$ Yes: R   No: W

_____ 6. $f(x) = \dfrac{1}{2}x - 6$ $g(x) = \dfrac{1}{2}x + 6$ Yes: S   No: O

_____ 7. $f(x) = 3x + 7$ $g(x) = \dfrac{x + 7}{3}$ Yes: E   No: A

_____ 8. $f(x) = 3x - 9$ $g(x) = \dfrac{1}{3}x + 3$ Yes: R   No: S

_____ 9. $f(x) = \dfrac{2x + 1}{3}$ $g(x) = \dfrac{3x - 1}{2}$ Yes: E   No: J

_____ 10. $f(x) = \dfrac{3x - 5}{6}$ $g(x) = \dfrac{6x - 5}{3}$ Yes: U   No: I

_____ 11. $f(x) = \dfrac{2x - 7}{9}$ $g(x) = \dfrac{9x + 7}{2}$ Yes: N   No: S

_____ 12. $f(x) = 7x - 9$ $g(x) = \dfrac{x - 9}{7}$ Yes: T   No: V

_____ 13. $f(x) = 3x + 10$ $g(x) = \dfrac{x - 10}{3}$ Yes: E   No: T

_____ 14. $f(x) = -x + 4$ $g(x) = -x + 4$ Yes: R   No: E

_____ 15. $f(x) = 2x - 11$ $g(x) = \dfrac{x - 11}{2}$ Yes: A   No: S

_____ 16. $f(x) = \dfrac{2x - 1}{3}$ $g(x) = \dfrac{3x + 1}{2}$ Yes: E   No: C

_____ 17. $f(x) = \dfrac{4x + 7}{5}$ $g(x) = \dfrac{5x - 7}{4}$ Yes: S   No: H

Enrichment
3.4 Order Up!

Find two values for *x* and for *y* for which each of these number sentences will be true. If a situation is impossible, indicate so.

1. $|x + y| > |x| + |y|$

_____

_____

2. $|x + y| \geq |x| + |y|$

_____

_____

3. $|x + y| < |x| + |y|$

_____

_____

4. $|x + y| \leq |x| + |y|$

_____

_____

5. $|x + y| = |x| + |y|$

_____

_____

6. $|x - y| > |x| + |y|$

_____

_____

7. $|x - y| \geq |x| - |y|$

_____

_____

8. $|x - y| < |x| - |y|$

_____

_____

9. $|x - y| \leq |x| - |y|$

_____

_____

10. $|x - y| = |x| - |y|$

_____

_____

11. $|x + y| = |x - y|$

_____

_____

12. $|x + y| < |x - y|$

_____

_____

13. $|x + y| \leq |x - y|$

_____

_____

14. $|x + y| > |x - y|$

_____

_____

15. $|x + y| \geq |x - y|$

_____

_____

16. $|xy| > |x| \, |y|$

_____

_____

17. $|xy| \geq |x| \, |y|$

_____

_____

18. $|xy| < |x| \, |y|$

_____

_____

19. $|xy| \leq |x| \, |y|$

_____

_____

20. $|xy| = |x| \, |y|$

_____

_____

21. $|x \div y| < |x| \div |y|$

_____

_____

22. $|x \div y| > |x| \div |y|$

_____

_____

23. $|x \div y| \geq |x| \div |y|$

_____

_____

24. $|x \div y| \leq |x| \div |y|$

_____

_____

Enrichment
3.5 Piecewise Functions

A piecewise function is similar to a step function in that it is
defined over intervals of *x*. But unlike step functions, not all
sections of the graph are horizontal lines. Graph each piecewise
function on the grid provided.

1. $f(x) = \begin{cases} -x, & \text{if } -2 < x \leq 0 \\ 3, & \text{if } 0 < x < 3 \\ 2x, & \text{if } 3 \leq x < 6 \end{cases}$

2. $f(x) = \begin{cases} 4, & \text{if } 0 \leq x \leq 2 \\ 2x, & \text{if } 2 < x < 5 \\ -x, & \text{if } x \leq 5 \end{cases}$

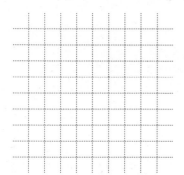
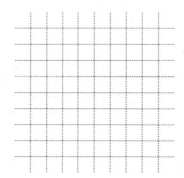

3. $f(x) = \begin{cases} x - 4, & \text{if } -3 \leq x \leq 1 \\ 2x, & \text{if } 1 < x < 4 \\ 5, & \text{if } x \geq 4 \end{cases}$

4. $f(x) = \begin{cases} -2x, & \text{if } -4 \leq x \leq 1 \\ 5, & \text{if } 1 < x < 3 \\ x + 1, & \text{if } x \geq 3 \end{cases}$

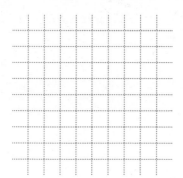
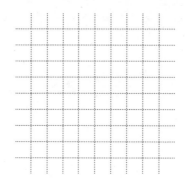

5. $f(x) = \begin{cases} -2x, & \text{if } -3 < x < -1 \\ 4, & \text{if } -1 \leq x \leq 2 \\ x + 4, & \text{if } x > 2 \end{cases}$

6. $f(x) = \begin{cases} x - 5, & \text{if } -3 \leq x \leq -1 \\ x + 1, & \text{if } -1 < x \leq 3 \\ 2x - 4, & \text{if } x > 3 \end{cases}$

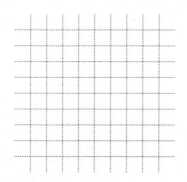
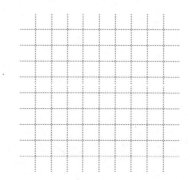

Enrichment
3.6 Parametric Equations and Substitution

A system of parametric equations can be written in the form $y = mx + b$ by using substitution and any method to eliminate t. For example, in the system $x(t) = 2t + 1$, you can write $x(t)$, then solve each equation for t.
$y(t) = t - 8$

$$x = 2t + 1 \qquad y = t - 8$$
$$x - 1 = 2t \qquad y + 8 = t$$
$$\frac{x - 1}{2} = t$$

Because $t = t$, $\qquad \dfrac{x - 1}{2} = y + 8$

$$\frac{1}{2}x - \frac{1}{2} - 8 = y$$
$$y = \frac{1}{2}x - \frac{17}{2}$$

Write each system in the form $y = mx + b$.

1. $\begin{cases} x(t) = t - 3 \\ y(t) = t + 1 \end{cases}$

2. $\begin{cases} x(t) = 2t - 1 \\ y(t) = t + 4 \end{cases}$

3. $\begin{cases} x(t) = 3t + 5 \\ y(t) = -2t + 1 \end{cases}$

_____

_____

_____

4. $\begin{cases} x(t) = -\frac{1}{2}t + 6 \\ y(t) = 3t - 4 \end{cases}$

5. $\begin{cases} x(t) = \frac{2}{3}t - 1 \\ y(t) = \frac{3}{4}t + 1 \end{cases}$

6. $\begin{cases} x(t) = -3t + 8 \\ y(t) = 4t - 5 \end{cases}$

_____

_____

_____

7. $\begin{cases} x(t) = 3t + 7 \\ y(t) = 2t - 10 \end{cases}$

8. $\begin{cases} x(t) = 3t - 4 \\ y(t) = \frac{1}{2}t + 1 \end{cases}$

9. $\begin{cases} x(t) = \dfrac{2t - 1}{3} \\ y(t) = -\frac{3}{4}t + \frac{1}{4} \end{cases}$

_____

_____

_____

Technology
3.1 Line and Point Symmetry

A geometric figure in the plane has line symmetry if, when the figure is folded along a line, the figure lies on top of itself. A geometric figure in the plane has point symmetry, if when it is rotated 180° around a point, the figure lies on top of itself. These definitions are illustrated in the diagrams.

The graph of an equation in one variable defines a set of points in the plane that may have symmetry. You can use a graphics calculator to explore symmetry and the trace feature to determine any line or point of symmetry. Be sure to choose an interval for x that is long enough to give a characteristic picture of the graph in question.

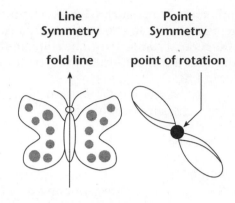

Line Symmetry   Point Symmetry

fold line   point of rotation

Use a graphics calculator to graph each equation. Then, if the graph has line symmetry, use the trace feature to find an equation of the line of symmetry or, if the graph has point symmetry, the coordinates of the point of rotation.

1. $f(x) = x^2$

2. $g(x) = x^2 - 2$

3. $h(x) = x^2 + 2x$

_____   _____   _____

4. $f(x) = -x^2 + 4x + 1$

5. $m(x) = x^2 - 6x$

6. $n(x) = -2x^2 - 4x + 1$

_____   _____   _____

7. $f(x) = x^3 - x^2$

8. $t(x) = x^3 + 3$

9. $f(x) = 2x^3 - 3x^2 + x$

_____   _____   _____

10. Explain why the graph of a function defined by an equation cannot be symmetric with respect to the x-axis.

_____

11. Suppose that $f(x) = ax^2 + bx + c$, where $a \neq 0$. Use a graphics calculator and experiment with different values of a, b, and c to find an equation whose graph is symmetric about the line $x = 4$.

_____

_____

Technology
3.2 Conjectures About Inverses

A conjecture is an educated guess about the truth of some statement. By exploring specific instances, you can formulate or test general statements. The statement shown is an example of a conjecture about linear functions whose truth you can test.

    If $f(x) = mx + b$, then f has an inverse.
The graphics calculator display shows the graphs of $f(x) = 2x + 1$, $g(x) = 1$, and $h(x) = -0.5x + 1$. The graph of f has a positive slope and passes the horizontal line test. The graph of g is horizontal and does not pass the horizontal line test. The graph of h has a negative slope and thus passes the horizontal line test. The conjecture is not always true. A truthful statement is, If $f(x) = mx + b$ and $m \neq 0$, then f has an inverse.

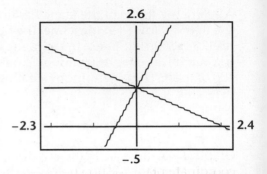

Use a graphics calculator to test the truth of each statement. If the statement is true, give a reasonable argument to show its truth. If the statement is sometimes true, correct the statement so that it is always true. If the statement is false, show how you know it is false.

1. The function $f(x) = x$ is its own inverse.

_____

2. If $f(x) = ax^2$ and $a > 0$, then f has an inverse.

_____

3. If $f(x) = ax^2$ and $a < 0$, then f has an inverse.

_____

4. If $f(x) = ax^3$, then f has an inverse.

_____

5. If $f(x) = \frac{a}{x}$ and $a \neq 0$, then f is its own inverse.

_____

6. The function $f(x) = ax^2$ has an inverse if $x \geq 0$.

_____

7. The function $f(x) = ax^2$ has an inverse if $x \leq 0$.

_____

8. If $m \neq 0$, then the inverse of $f(x) = mx + b$ is $g(x) = mx + b$.

_____

Technology
3.3 Simple Power Functions and Composition

A power function is any function of the form $f(x) = ax^n$, where $a \neq 0$ and n is a positive integer. If you let $a = 1$, you get simple power functions.

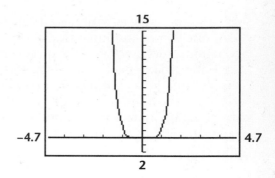

You can pose and answer some questions about the properties of the set of simple power functions under composition by using your graphics calculator. For example, If $f_2(x) = x^2$ and $f_3(x) = x^3$, does $(f_2 \circ f_3)(x) = (f_3 \circ f_2)(x)$ for each real number x?

Graph $(f_2 \circ f_3)(x) = (x^3)^2$ and $(f_3 \circ f_2)(x) = (x^2)^3$ on the same coordinate plane as shown. Notice that the graphs coincide. Thus, composition of f_2 with f_3 is commutative.

Use a graphics calculator to answer each question.

1. If $f_4(x) = x^4$ and $f_3(x) = x^3$, does $(f_4 \circ f_3)(x) = (f_3 \circ f_4)(x)$? _____

2. If $f_1(x) = x^1$ and $f_5(x) = x^5$, does $(f_1 \circ f_5)(x) = (f_5 \circ f_1)(x)$? _____

3. Do you think that $(f_m \circ f_n)(x) = (f_n \circ f_m)(x)$ for all positive integers m and n?

   _____

4. Do you think the composition of two simple power functions is another simple power function?

   _____

5. Describe the graph of $(f_m \circ f_n)(x)$ if both m and n are even.

   _____

6. Describe the graph of $(f_m \circ f_n)(x)$ if both m and n are odd.

   _____

7. If $f(x) = 2x - 3$ and $g(x) = 3x + 1$, graph $f \circ g$ and $g \circ f$ on the same coordinate plane.

8. If $f(x) = 0.5x + 1$ and $g(x) = 2x + 3$, graph $f \circ g$ and $g \circ f$ on the same coordinate plane.

9. Do you think the composition of two linear functions is also linear?

   _____

10. Show by example that composition of linear functions is not commutative.

   _____

   _____

Technology

3.4 Solving Absolute Value Equations

An absolute value equation is an equation that involves at least one pair of absolute value symbols. Some absolute value equations, such as $|x| = 3$, are quite simple. Others, such as $|2x + 3| = -|3x - 4| + 7$, are quite involved. You can solve simple equations with paper and pencil or even mentally. You can solve the more complicated ones by viewing each side of the equation as defining a function and then using a graphics calculator. The diagram shows the graphs of $f(x) = |2x + 3|$ and $g(x) = -|3x - 4| + 7$ on the same coordinate plane.

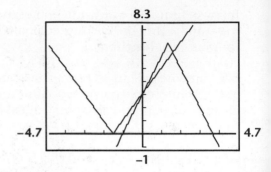

The calculator display shows that one solution is $x = 0$. You can use the trace feature to find that the second solution is $x = 1.6$.

Use a graphics calculator to solve each absolute value equation. If no solution exists, write "no solution."

1. $|2x + 2| = 6$

_____

2. $|x| = 2|x|$

_____

3. $|3x - 1| = 0$

_____

4. $-3 = |4x + 5|$

_____

5. $|3x + 10| = -|x - 5|$

_____

6. $|x - 1| + 5 = |3x| + |3x - 1|$

_____

Use a graphics calculator to solve each absolute value inequality. If no solution exists, write "no solution."

7. $|2x + 1| \geq 0$

_____

8. $|x| \leq x$

_____

9. $|7x - 1| > 4$

_____

10. $|-5x + 1| < |3x + 1|$

_____

Technology
3.5 Variations on the Greatest Integer Function

Interesting things happen when you decide to compose functions. What would happen, for example, if you compose $f(x) = [x]$ with $g(x) = 0.5x^2$ to form $(f \circ g)(x) = [0.5x^2]$? To find out, you can use a spreadsheet or a graphics calculator. The bar graph shows a chart generated by using a spreadsheet and the following values for x.

$$x = -5, -4.5, -4, ..., 4, 4.5, 5$$

The values of $(f \circ g)(x)$ are generated by using the rounding function with zero decimal places.

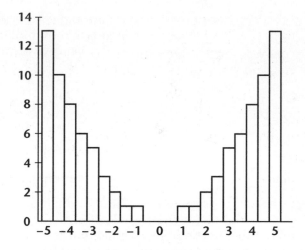

Use a spreadsheet or graphics calculator to graph each composition.

1. $f(x) = [x]; g(x) = 0.5x; f \circ g$

2. $f(x) = [x]; g(x) = 2x; f \circ g$

3. $f(x) = [2x]; g(x) = 0.5x; f \circ g$

4. $f(x) = [2x]; g(x) = 0.5x; g \circ f$

5. $f(x) = [x]; g(x) = x^3; f \circ g$

6. $f(x) = [x]; g(x) = x^3; g \circ f$

7. $f(x) = 2[x - 1]; g(x) = 2x - 1; f \circ g$

8. $f(x) = [x]; f \circ f$

9. $f(x) = [x]^2; g(x) = x^2; f \circ g$

10. $f(x) = [x]; g(x) = [0.5x]; h(x) = [0.25x]; g \circ (f \circ h)$

11. Describe the composition of $f(x) = [x]$ with itself n times, where n is a positive integer.

_____

Technology
3.6 Division Points of Line Segments

If you are given two points on the coordinate plane, say points $P(a, b)$ and $Q(r, s)$, you can use parametric equations to find the coordinates of any point along the line segment $\overline{PQ}$. The following parametric equations give the x- and y-coordinates of any point F on $\overline{PQ}$.

$$x = (1 - t)a + tr \qquad y = (1 - t)b + ts$$

The variable t satisfies the inequality $0 \le t \le 1$.

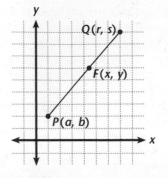

With a spreadsheet, you can find the coordinates of the points that divide the given line segment into any number of congruent segments. The spreadsheet shows the division of points of the segment with endpoints $P(0, 1)$ and $Q(6, 4)$.

Notice that if $t = 0.5$, $x = 3.0$, and $y = 2.5$. These values are the averages of the x-coordinates and y-coordinates of P and Q, respectively.

	A	B	C
1	T	X	Y
2	0.00	0.0	1.00
3	0.25	1.5	1.75
4	0.50	3.0	2.50
5	0.75	4.5	3.25
6	1.00	6.0	4.00

Cell B2 contains
$(1-A2)*0+A2*6$.
Cell C2 contains
$(1-A2)*1+A2*4$.

Create a spreadsheet that gives the coordinates of the requested division points of the line segment with first endpoint P and second endpoint Q.

1. $P(-4, -4)$; $Q(4, 4)$; two congruent segments _____

2. $P(0, 0)$; $Q(4, 4)$; seven congruent segments _____

3. $P(-4, -4)$; $Q(0, 4)$; two congruent segments _____

4. $P(0, 0)$; $Q(10, 0)$; four congruent segments _____

5. $P(0, 0)$; $Q(0, 6)$; six congruent segments _____

6. $P(-4, -2)$; $Q(10, 4)$; six congruent segments _____

7. $P(-6, 6)$; $Q(6, -6)$; four congruent segments _____

8. $P(1.5, 0)$; $Q(0, 5)$; six congruent segments _____

9. $P(0, 5)$; $Q(15, 0)$; six congruent segments _____

10. Find the coordinates of point R on the line segment whose endpoints are $P(2, 3)$ and $Q(7, 5)$ if $\dfrac{PR}{RQ} = \dfrac{2}{15}$.

Lesson Activity
3.1 Symmetry Patterns

Hexagons and other geometric figures can be used alone or together to form repeating patterns. One such pattern is shown here.

Draw a set of coordinate axes through the center of the pattern. Then follow the directions.

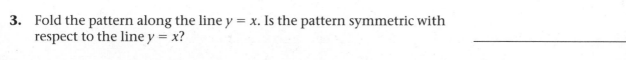

1. Fold the pattern along the x-axis. Is the pattern symmetric with respect to the x-axis?

   _____

2. Fold the pattern along the y-axis. Is the pattern symmetric with respect to the y-axis?

   _____

3. Fold the pattern along the line $y = x$. Is the pattern symmetric with respect to the line $y = x$?   _____

4. Draw another line of symmetry. Write the equation of the line you

   drew. _____

5. In the space provided, create your own geometric design.

6. Examine the design. Then name the symmetries.

   _____

   _____

Lesson Activity
3.2 Exploring Inverse Functions Using a Graphics Calculator

To graph $f(x) = 2x + 5$ and its inverse, $f^{-1}(x)$, along with $y = x$ in parametric representation, graph the respective systems:

$$x(t) = t \qquad\qquad x(t) = 2t + 5 \qquad x(t) = t$$
$$y(t) = 2t + 5 \qquad y(t) = t \qquad\qquad y(t) = t$$

With your calculator in parametric mode, enter $X_{1T} = T$, $Y_{1T} = 2T + 5$, $X_{2T} = 5$, $Y_{2T} = T$, $X_{3T} = T$, and $Y_{3T} = T$.

Set the viewing window at Tmin $= -20$, Tmax $= 20$, and Tstep $= 1$ (or any other similar viewing window).

1. Graph the parametric equations.

2. Complete the chart.

	$f(x)$		$f^{-1}(x)$	
t	x	y	x	y
-10				
-5				
0				
5				
10				

3. When t is -10, what is $f(-10)$? _____ $f^{-1}(-10)$? _____

4. Where do f and f^{-1} intersect? _____

5. If (a, b) is a point on the graph of f, what is $f(a)$? _____ $f^{-1}(b)$? _____

6. Explain why f^{-1} is a function. _____

7. Write the linear equation for f^{-1}. _____

Graph $g(x) = x^2 + 2$ and its inverse, using parametric representation.

8. Does g^{-1} exist? Explain. _____

9. How can you restrict the parametric "Range" on the graphics calculator so that g^{-1} exists?

_____

Lesson Activity
3.3 Decision Making

Amanda works part time as a clerk each day after school. She must make a choice about working after school or studying for an Advanced Algebra test. She can choose from no studying and all work to no work and all studying, with various choices between these two. The table lists five different combinations of income earned and possible test scores based on five different ways she can proportion her time.

Choice of time	Income Earned	Test Score
0 hours working 4 hours studying	0	95
1 hour working 3 hours studying	6	85
2 hours working 2 hours studying	12	75
3 hours working 1 hour studying	18	65
4 hours working 0 hours studying	24	55

1. Write the function f using Income Earned as the domain and Test Score

   as the range. _____

2. Write the function g using Hours Working as the domain and Income

   Earned as the range. _____

3. Write the composition of f and g. _____

4. What is the domain and range of the composition function?

   _____

5. Graph the composition function using a graphics calculator.

6. Find Amanda's test score if she works 1.5 hours. _____

7. Suppose Amanda must work an average of at least 3 hours a day and she wants a test score of at least 80. Is this possible? What are some alternatives?

   _____

Lesson Activity
3.4 The Absolute Value Function

Each of the following graphs is the graph of an absolute value function.

**Use a computer or graphics calculator to find a function rule to
match each graph. Record the function rules used, as well as the
final answer. Explain how the trial functions were used to
arrive at the final answer.**

1.

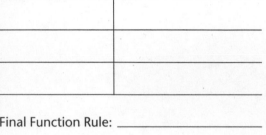

Trial Function Rule	General Shape of Graph

Final Function Rule: _____

Reasoning: _____

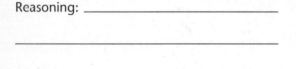

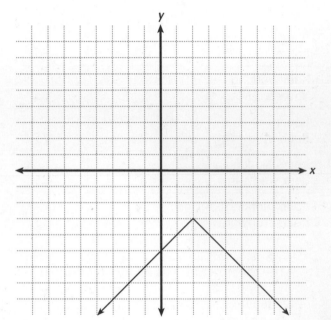

2.

Trial Function Rule	General Shape of Graph

Final Function Rule: _____

Reasoning: _____

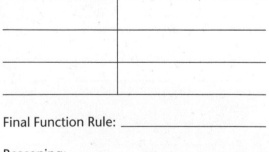

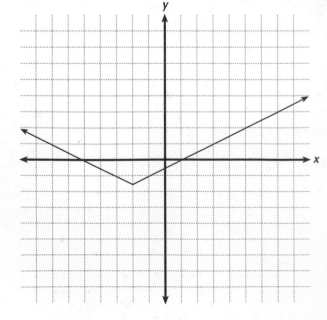

Lesson Activity
3.5 Hardiness Zones

The U.S. Department of Agriculture divides the map of the
United States into hardiness zones based on average annual
temperatures for each zone. Examine the table showing the
approximate range of temperatures in degrees Fahrenheit for
each hardiness zone.

Temperature Range (°F)	Zone
$-40 \leq x < -30$	3
$-30 \leq x < -20$	4
$-20 \leq x < -10$	5
$-10 \leq x < 0$	6
$0 \leq x < 10$	7
$10 \leq x < 20$	8
$20 \leq x < 30$	9
$30 \leq x < 40$	10
$40 \leq x < 50$	11

1. Graph the temperature range vs. zone on
   the grid provided.

2. What type of step function is represented?

   _____

3. Write an equation for the step function
   that describes the hardiness zone data.

   _____

4. What is the hardiness zone for your area
   of the United States?

   _____

5. Create your own step function. Write the equation and
   graph it on the grid provided.

   _____

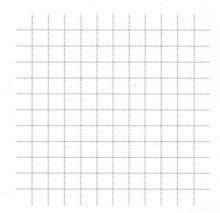

Lesson Activity
3.6 Exploring Parametric Equations

Graph these parametric systems using your graphics calculator in
parametric mode and simultaneous format. Use the viewing window
Tmin = −20, Tmax = 30, Tstep = 0.5, Xmin = −20, Xmax = 30, Xscl = 5,
Ymin = −20, and Ymax = 40 (or any similar viewing window.

$$\begin{cases} x(t) = 2t - 8 \\ y(t) = t \end{cases} \qquad \begin{cases} x(t) = t \\ y(t) = t \end{cases} \qquad \begin{cases} x(t) = t \\ y(t) = 2t - 8 \end{cases}$$

1. Describe what you notice as the parametric representations are graphed.

   _____

2. Experiment with the trace feature and the up- and down-arrow keys.
   What relationship between the graphs is shown?

   _____

3. Write the function described by the first set of parametric equations.

   _____

4. Write the inverse of that function. _____

5. State the range and domain of the first function. _____

6. State the range and domain of the inverse function. _____

7. Is the inverse a function? Explain. _____

   _____

**Graph $g(x) = |2x - 8|$ and its inverse using parametric
representation.**

8. Is the inverse of $g(x)$ a function? _____

Graph $h(x) = x^2 + 4x + 5$.

9. How could you restrict the domain of $h(x)$ so that its inverse is a
   function?

   _____

Assessing Prior Knowledge
3.1 Exploring Symmetry

Plot the following points on the grid provided.

1. $A(2, 4)$

2. $B(-2, 4)$

3. $C(4, 2)$

4. $D(-4, -2)$

Which two of the above points, if any, are the same distance from

5. the x-axis? _____

6. the y-axis? _____

- -

Quiz
3.1 Exploring Symmetry

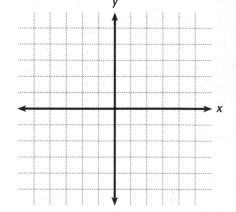

1. Draw the triangle with vertices $A(-3, -4)$, $B(-1, 2)$, and $C(5, 0)$ on the grid provided. Determine the image vertices A', B', and C' that are symmetric to A, B, and C with respect to the y-axis. Plot these image points and draw $\triangle A'B'C'$.

_____

2. Determine the coordinates of the image vertices that are symmetric to vertices of $\triangle ABC$ with respect to the x-axis.

_____

3. Complete the chart.

Pre-image point	Image point	Symmetric with Respect to
$(3, -2)$	$(-2, 3)$	_____
_____	$(7, 9)$	y-axis
$(-10, 2)$	_____	$y = x$
_____	$(0, 0)$	x-axis
$(8, 10)$	_____	x-axis
$(11, -12)$	$(11, 12)$	_____

Assessing Prior Knowledge
3.2 Inverse Functions

Plot the set of ordered pairs. Then plot the image points that are symmetric to the set with respect to the line $y = x$.

1. $\{(1, -3), (2, 2), (3, 2)\}$

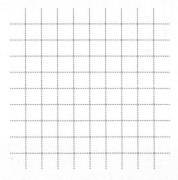

- -

Quiz
3.2 Inverse Functions

1. Find the inverse of the function $f = \{(-2, -8), (-1, -1), (3, 7), (3, 27)\}$.

_____

Determine whether each function has an inverse function. Draw a rough sketch of the inverse if it exists; otherwise write no inverse.

2.

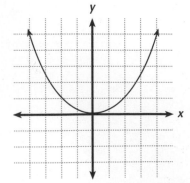

3.

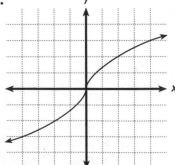

4.

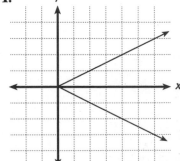

Find the inverse of each function.

5. $f(x) = 2\left(x + \frac{1}{2}\right)$

6. $f(x) = \frac{2}{3}x + 3$

7. $f(x) = \frac{1}{x - 1}, x \neq 1$

_____   _____   _____

Assessing Prior Knowledge
3.3 Composition of Functions

Evaluate each function for $x = -1$, $x = 2$, and $x = 0$.

1. $g(x) = x^2 - 2x$ _____

2. $h(x) = \dfrac{x + 1}{2}$ _____

- -

Quiz
3.3 Composition of Functions

For $f(x) = 2x + 4$ and $g(x) = x^2 + 1$, find each of the following.

1. $f \circ g$ **2.** $g \circ f$ **3.** $(f \circ g)(2)$

_____   _____   _____

4. $(g \circ f)(-1)$ **5.** $(f \circ f)(-3)$ **6.** $(g \circ f)(0)$

_____   _____   _____

7. Let $f(x) = -2x$ and $g(x) = x + \dfrac{1}{2}$. Find $f \circ g$ and $g \circ f$ and graph each on the grid provided.

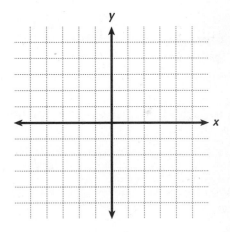

8. Find $g \circ (f \circ g)$ for f and g in Exercise 7.

_____

Mid-Chapter Assessment
Chapter 3 (Lessons 3.1 – 3.3)

Write the letter that best answers the question or completes the statement.

_____ **1.** Given the set of points {(1, 2), (9, 10), (11, 12)}, which image set is symmetric with respect to the x-axis?

    a. {(−2, 1), (−10, −9), (−12, −11)}   **b.** {(2, 1), (10, 9), (12, 11)}
    c. {(1, −2), (9, −10), (11, −12)}   **d.** {(−1, 2), (−9, 10), (−11, 12)}

_____ **2.** Which of the following is the inverse of the function $h(x) = x^2 + 1$?

    a. $h^{-1}(x) = \sqrt{x^2 - 1}$ **b.** $h^{-1}(x) = \sqrt{x} - 1$
    c. $h^{-1}(x) = \sqrt{x - 1}$ **d.** $h^{-1}(x) = y^2 + 1$

_____ **3.** Given $f = \{-3, 2), (7, -9), (-0.5, 4)\}$, which represents f^{-1}?

    a. {(2, −3), (−9, 7), (4, −0.5)}   **b.** {(3, 2), (−7, −9), (0.5, 4)}
    c. {(−3, −2), (7, 9), (−0.5, 4)}   **d.** {−2, 3), (9, 7), (−4, 0.5)}

In Exercises 4–5, let $f(x) = 3x^2 - 2$ and $g(x) = \dfrac{2x}{5}$.

_____ **4.** Find $f \circ g$.

    a. $\dfrac{6x^2}{25} - \dfrac{4}{5}$ **b.** $\dfrac{12x^2}{25} - \dfrac{4}{5}$

    c. $\dfrac{6x^2}{5} - 2$ **d.** $\dfrac{12x^2}{25} - 2$

_____ **5.** Find $(g \circ f)(2)$.

    a. −4   **b.** 4   **c.** $\dfrac{-2}{25}$ **d.** 5

6. Find the inverse of the function $g(x) = \dfrac{-3}{2}x - 4$. _____

7. Use the grid provided to reflect quadrilateral $ABCD$ over the line $y = x$. List the image points A', B', C', and D'.

_____

_____

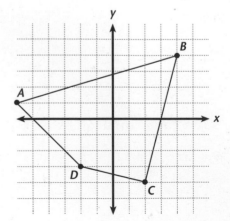

Assessing Prior Knowledge
3.4 The Absolute Value Function

Graph each function on the same coordinate plane.

1. $f(x) = 2x + 1$

2. $g(x) = -2x - 1$

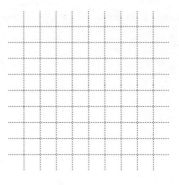

- -

NAME _____ CLASS _____ DATE _____

 Quiz
3.4 The Absolute Value Function

Simplify.

1. $|-4| - |-5|$ _____

2. $|-2.5| + |-1.5|$ _____

3. $-|3| - |-3|$ _____

For Exercises 4–6, write the function that is graphed.

4.

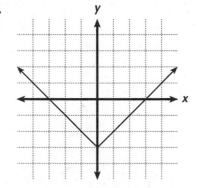

5.

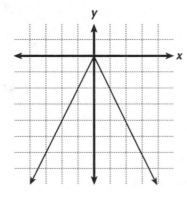

6.

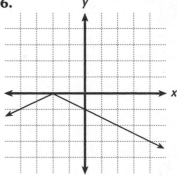

_____

_____

_____

Solve each equation using the definition of absolute value.

7. $|3x - 2| = 11$

8. $|x + 1| + 2 = 7$

9. $|-2x| + 1 = 8$

_____

_____

_____

Assessing Prior Knowledge
3.5 Step Functions

Graph each inequality on a number line.

1. $-2 < x \le 0$

2. $4 \le x < 7.5$

Quiz
3.5 Step Functions

Simplify.

1. $[-4.1] + [3.2]$ _____

2. $[-0.5] - [-1.3]$ _____

3. $[-\sqrt{2}] + [1.5]$ _____

4. $[7.99] + [0.51] - [-3.98]$ _____

5. $\lceil 0.5 \rceil - \lceil 1.7 \rceil$ _____

6. $\lceil 1.67 \rceil - \lceil 0.59 \rceil$ _____

7. $\lceil -6.759 \rceil + \lceil 2 \rceil$ _____

8. $\lceil 0.75 \rceil + \lceil 3.14 \rceil$ _____

Pete's Plumbing charges a $75 base fee for the first hour of work and $15 for every additional hour or fraction of an hour.

9. Write the function that relates Pete's charges to time.

_____

10. Use your own graph paper to graph the function you found in Exercise 9 over the interval $0 \le t \le 4$.

11. How much does Pete charge for a job that takes 3.75 hours? _____

Assessing Prior Knowledge
3.6 Parametric Equations

Write an equation for each line described.

1. slope -2, contains $(-4, -3)$ _____

2. contains $(-4, 2)$ and $(6, 7)$ _____

3. contains $(5, 2)$ and $(3, 2)$ _____

- -

Quiz
3.6 Parametric Equations

1. Graph the following parametric system for $0 \le t \le 4$ on the grid provided.

$$\begin{cases} x(t) = 3t \\ y(t) = 2t - 1 \end{cases}$$

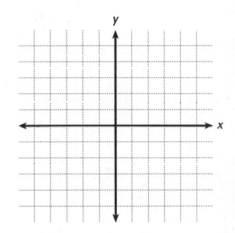

Combine each system of parametric equations into one linear function.

2. $\begin{cases} x(t) = 3t \\ y(t) = \frac{1}{2}t + 1 \end{cases}$

3. $\begin{cases} x(t) = -t \\ y(t) = -4t - 2 \end{cases}$

4. $\begin{cases} x(t) = \frac{t}{3} \\ y(t) = 5t \end{cases}$

_____   _____   _____

An icemaker produces uniform spherical ice cubes with radii of 15 mm. The cross-sectional circumference of each "cube" is approximately 94.2 mm. At 15°C the radius of the ice cube decreases 1 mm per minute and the cross-sectional circumference decreases 6.28 mm per minute.

5. Write the parametric equations that describe the situation for t minutes.

_____

6. What will the cross-sectional circumference be at $t = 15$ min?

_____

Chapter Assessment
Chapter 3, Form A, page 1

Write the letter that best answers the question or completes the statement.

1. Given the pre-image set of ordered pairs {(0, 1), (1, 3), (−1, −1)} and the image set {−1, 0), (−3, −1), (1, 1)}, with respect to which line are they symmetric?

    a. x-axis   **b.** y-axis   **c.** $y = x$ **d.** $y = -x$

2. Identify the image point that is symmetric to the pre-image point (x, y) with respect to the y-axis.

    a. (y, x) **b.** $(x, -y)$ **c.** $(-x, y)$ **d.** $(-y, x)$

3. Which is the inverse of f = {(−2, 4), (1, 2), (0, 0), (4, −8)}?

    a. {(2, 4), (−1, −2), (0, 0), (−4, −8)}
    b. {(4, −2), (−2, 1), (0, 0), (−8, 4)}
    c. {(2, −4), (−1, 2), (0, 0), (−4, 8)}
    d. {(−4, 2), (2, −1), (0, 0), (8, −4)}

4. Which function is the equation for the graph shown?

    a. $f(x) = 2|x|$

    b. $f(x) = \frac{1}{2}|x|$

    c. $f(x) = -\frac{1}{3}|x|$

    d. $f(x) = 3|x|$

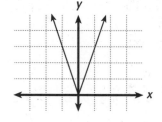

5. Which are the solutions to the equation $|4x - 3| = 8$?

    a. $x = -1\frac{1}{4}$ and $x = 2\frac{3}{4}$ **b.** $x = -2\frac{3}{4}$ and $x = 1\frac{1}{4}$

    c. $x = -1\frac{1}{4}$ and $x = -2\frac{3}{4}$ **d.** $x = 1\frac{1}{4}$ and $x = 2\frac{3}{4}$

6. When simplified, $\lceil -7.5 \rceil + -[0.5] - [3.2] =$

    a. 10.5   **b.** −5   **c.** −6.5   **d.** −11

7. The value of $\lceil 9.3 \rceil - [5.67]$ is

    a. 5   **b.** 4   **c.** 3   **d.** 3.62

Chapter Assessment
Chapter 3, Form A, page 2

_____ **8.** If $f(x) = 2x - 1$ and $g(x) = \frac{x^2}{2}$, which of the following represents $f \circ g$?

a. $x^2 - 1$ **b.** $2x^2 - 2x + \frac{1}{2}$

c. $x^3 - \frac{x^2}{2}$ **d.** $\frac{x^2}{2} + 2x - 1$

_____ **9.** If $g(x) = -3x^2$ and $h(x) = \frac{4}{3}x$, which of the following represents $(h \circ g)(2)$?

a. $\frac{8}{3}$ **b.** $\frac{-64}{3}$ **c.** -12 **d.** -16

_____ **10.** The inverse of $h(x) = \frac{4}{3}x^2 + 3$ is

a. $h^{-1}(x) = \frac{3x - 9}{4}$ **b.** $h^{-1}(x) = \frac{1}{2}\sqrt{3x - 3}$

c. $h^{-1}(x) = \frac{1}{2}\sqrt{3x - 9}$ **d.** $h^{-1}(x) = \frac{3}{4}\sqrt{x - 3}$

_____ **11.** If $f(x) = \sqrt{3x + 2}$, which of the following are the domain and range of $f(x)$?

a. Domain: $x \geq -\frac{2}{3}$
Range: All real numbers

b. Domain: $x \geq -\frac{2}{3}$
Range: $y \geq 0$

c. Domain: $x \geq 0$
Range: $y \geq 0$

d. Domain: $x \geq 0$
Range: All real numbers

_____ **12.** Which function represents the parametric equations

$$\begin{cases} x(t) = \frac{-3}{2}t \\ y(t) = 2t + 2? \end{cases}$$

a. $f(x) = -3x + 2$ **b.** $f(x) = -\frac{4}{3}x - 2$

c. $f(x) = -\frac{3}{4}x - 2$ **d.** $f(x) = -\frac{4}{3}x + 2$

Chapter Assessment
Chapter 3, Form B, page 1

1. Given the set of ordered pairs $\{(-2, \frac{2}{3}), (1, -\frac{1}{3}), (3, -1)\}$, list the symmetric image set of ordered pairs with respect to the line $y = x$.

    _____

2. List the image points that are symmetric to the vertices of $\triangle ABC$ with respect to the x-axis.

    _____

    _____

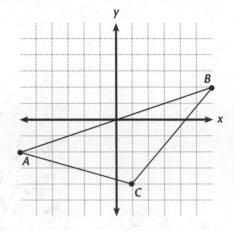

3. Given the function $\{(0, 0.5), (3, -14.5), (1, -4.5), (-1, 5.5)\}$, find the inverse function.

    _____

Give the range for each function.

4. $g(x) = 4|x| + 1$ 5. $h(x) = -\frac{1}{2}|x|$ 6. $f(x) = |2x| - 3$

    _____   _____   _____

7. Solve the equation: $|7x - 3| + 2 = 12.$ _____

Simplify.

8. $3|-1.9| - |-10| =$ _____   9. $[8.25] + [7.3] - [0.5] =$ _____

10. $\lceil 3.2 \rceil + \lceil -7.69 \rceil =$ _____   11. $|4.1| - \lceil 3.73 \rceil + [-1.59] =$ _____

For $f(x) = x^2$, $g(x) = x - 2$, **and** $h(x) = \sqrt{2x + 1}$, **find:**

12. $g \circ f$ _____

13. $(g \circ f)(-1)$ _____

14. $f \circ h$ _____

15. $(h \circ f)(0)$ _____

16. $h \circ g$ _____

17. $(f \circ g)(3)$ _____

18. Use the grid provided to graph $f(x) = \dfrac{x + 4}{5}$ and $g(x) = 5x - 4$ on the same set of axes.

19. Using composition, show that the functions in Exercise 18 are inverse functions.

_____

_____

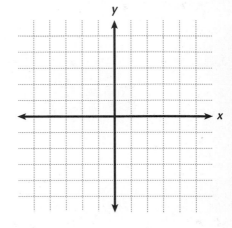

20. Combine the parametric equations $\begin{cases} x(t) = \frac{1}{2}t + 1 \\ y(t) = 5t \end{cases}$

to make one linear function

_____

Mike is filling a spherical water balloon. Each second the volume of the balloon increases 100 mL as the diameter expands approximately 1.8 cm.

21. Write the parametric equations to describe the situation for t seconds.

_____

22. What will be the diameter of the balloon at $t = 8$ seconds?

_____

Alternative Assessment
Inverses and Composition of Functions, Chapter 3, Form A

TASK: To find and recognize inverse functions and find composite functions

HOW YOU WILL BE SCORED: As you work through the task, your teacher will be looking for the following:

- how well you understand the definition of the inverse of a function
- whether you can describe the relationship between the graphs of a function and its inverse.
- how effectively you can describe what results from the composition of any invertible function with its inverse function.

Let $f(x) = 3x + 2$ and $g(x) = \frac{1}{3}x - \frac{2}{3}$.

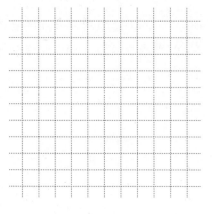

1. Graph each function.

2. Find the domain and the range of both f and g. Describe what you notice about the domain of f and the range of g. Describe what you notice about the range of f and the domain of g.

   _____

   _____

3. Are f and g inverse functions? Describe how you can determine whether or not f and g are inverse functions by examining their graphs?

   _____

   _____

4. What is the equation of the axis of symmetry of the two graphs? _____

5. Find $f \circ g$ and $g \circ f$. How do $f \circ g$ and $g \circ f$ compare with the equation of the axis of symmetry?

   _____

6. What is the composition of any function and its inverse? _____

SELF-ASSESSMENT: What does it mean when a function has an inverse relation but not an inverse function?

_____

Alternative Assessment
Applications of Functions, Chapter 3, Form B

TASK: To solve a real-world problem using functions

HOW YOU WILL BE SCORED: As you work through the task, your teacher will be looking for the following:

- whether you can determine the function represented by a real-world problem
- how well you solve the problem

The New Jersey Transit bus fare is based on a chart of zones showing the fares for travel between zones. This chart relates zones to the cost of a bus ticket.

Zone	Bus Fare
$0 \leq z < 1$	$0.75
$0 \leq z < 2$	1.50
$0 \leq z < 3$	2.25
$0 \leq z < 4$	3.00
$0 \leq z < 5$	3.75
$0 \leq z < 6$	4.50
$0 \leq z < 7$	5.25
$0 \leq z < 8$	6.00
$0 \leq z < 9$	6.75
$0 \leq z < 10$	7.50

1. Graph the data in the table.

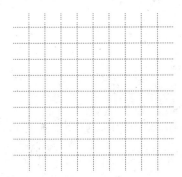

2. Discuss why a step function represents this data.

   _____

3. Is this step function for bus fare a greatest-integer function or a rounding-up function? Explain.

   _____

4. Write a step function that can be used to find the bus fare based on the zone.

   _____

5. Describe the type of transformation represented in the bus fare graph. Explain, in general terms, how the step function is transformed.

   _____

   _____

   _____

SELF-ASSESSMENT: Suppose the bus fare structure were not a function. Discuss the effect this would have on the cost of a bus ticket.

ANSWERS

Lesson 1.1

1. Yes; Constant difference between consecutive x- and y-values.

2. No

3. Yes; Constant difference between consecutive x- and y-values.

4. No 5. Yes 6. No 7. No 8. Yes

9. The y-values decrease by 3.

10. 2 11. $y = -3x + 2$

12. Consecutive x- and y-values are related by a constant difference.

13. D 14. C 15. A 16. B

Lesson 1.2

1. 0.2; -3 2. -4; 7 3. -1; 0 4. $\frac{1}{2}$; -6

5. 0; 5 6. $-\frac{3}{4}$; 8 7. $y = 3x + 4$

8. $y = -2x$ 9. $y = -\frac{1}{4}x + 5$ 10. $y = -6$

11. 2; $y = 2x - 1$ 12. -3; $y = -3x + 1$

13. 1; $y = x - \frac{7}{3}$ 14. $\frac{3}{2}$; $y = \frac{3}{2}x - \frac{1}{2}$

15. C 16. A

17. The answers are the same since the order of points does not change the slope.

18. Check students' graphs. 19. 93.97

Lesson 1.3

1.

2. positive correlation 3. (64, 40) 4. 0.94

5.

6. The points tend to slope downward.

7. -0.64 8. (25, 1.5) 9. $y = -0.05x + 3.7$

10. A scatter plot does not show a cause and effect relationship.

Lesson 1.4

1. $-\frac{3}{2}$; $y = -\frac{3}{2}x$ 2. $-\frac{3}{5}$; $y = -\frac{3}{5}x$

3. 2; $y = 2x$ 4. $C = kd$ 5. $I = kh$

6. The straight line passes through the point (0,0).

7. 2 8. $y = 2x$ 9. -4

10. The ratios are equal.

11. $-\frac{2}{5}$; $y = -\frac{2}{5}x$ 12. -1; $y = -x$

Lesson 1.5

1. 1 2. 9 3. 3 4. 2 5. -6 6. -2

7. 13.5 8. 5 9. -4 10. -1 11. -9

ANSWERS

12. $\frac{1}{2}$ **13.** 2.7 **14.** 0.5 **15.** -1.2

16. $20°, 70°$ **17.** 4.5 **18.** $-\frac{3}{2}$

19. $P = \frac{A}{(1 + rt)}$ **20.** $F = \frac{9}{5}C + 32$

21. $d = \frac{a_n - a_1}{n - 1}$ **22.** $h = \frac{2A}{b_1 + b_2}$

23. $-\frac{1}{3}$ **24.** -7 **25.** 4 **26.** 18 **27.** $150

28. 250 at $12; 150 at $9

Lesson 1.6

1. $x > 0$ **2.** $x < 5$

3. $x > 3$

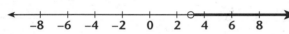

4. $x \geq 5$

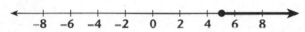

5. $x \geq -4\frac{1}{4}$

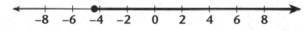

6. $x > 2$

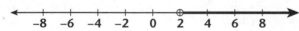

7. $x < 0$

8. $x \geq 2$

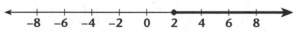

9. $y < -3x + 2$

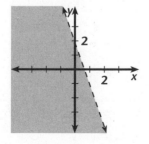

10. $y > \frac{3}{2}x + 1$

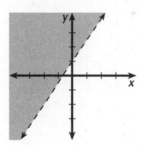

11. $y \leq \frac{1}{2}x - \frac{3}{2}$

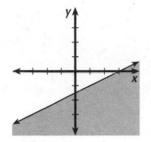

Enrichment — Chapter 1

Lesson 1.1

1.

Hours	Underwater	Waterworks
0	$ 60	$ 75
1	$116	$127
2	$172	$179
3	$228	$231
4	$284	$283
5	$340	$335
6	$396	$387
7	$452	$439
8	$508	$491
9	$564	$543
10	$620	$595

Underwater would charge $620; Waterworks would charge $595; Choose Waterworks for the job; Underwater: $y = 60 + 56x$; Waterworks $y = 75 + 52x$, where $x =$ number of hours worked.

ANSWERS

2.

Number of Students	Profit
0	−300
10	−270
20	−240
30	−210
40	−180
50	−150
60	−120
70	−90
80	−60
90	−30
100	0

100 students are needed to break even;
$y = 3x − 300$, where x = number of students.

3.

Weight (lb)	Cost
1.5	$1.30
2.0	$1.34
2.5	$1.38
3.0	$1.42
3.5	$1.46
4.0	$1.50
4.5	$1.54
5.0	$1.58

It would cost $1.58 to mail 5 lb of bound printed matter, $y = 1.30 + 0.04x$, where x = number of 0.5 lb increases over 1.5 lb.

4.

Month	Cost
0	$ 100
1	$ 180
2	$ 260
3	$ 340
4	$ 420
5	$ 500
6	$ 580
7	$ 660
8	$ 740
9	$ 820
10	$ 900
11	$ 980
12	$1060

It would cost $1060 the first year,
$y = 80x + 100$, where x = number of months.

Lesson 1.2

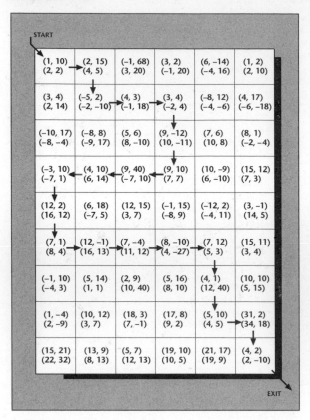

The slopes along the path through the maze are: $-8, -5, -4, -3, 0, 1, \frac{3}{2}, \frac{15}{8}, 2, \frac{9}{4}, \frac{5}{2}, 3, \frac{7}{2}, 4, \frac{17}{4}, \frac{9}{2}, \frac{39}{8}, 5, \frac{16}{3}$, and 6.

Lesson 1.3

1.

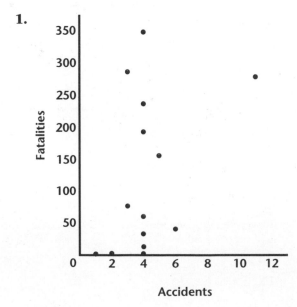

ANSWERS

2.

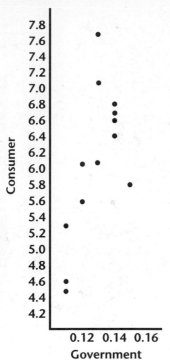

positive

3.

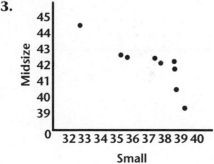

negative

4.

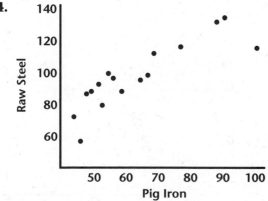

positive

5.

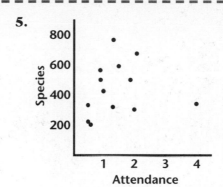

6.

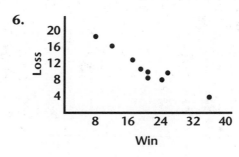

negative

7.

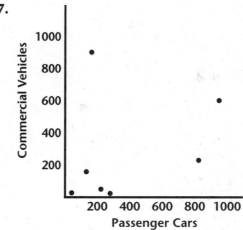

8. FITLINE

Lesson 1.4

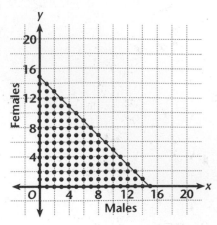

x	y
0	1
1	3
2	5

x	y
0	0
1	2
2	4

x	y
0	0
2	1
4	2

x	y
2	−4
4	−8
6	−12

x	y
4	1
8	3
12	5

x	y
0	4
1	7
2	10

x	y
0	0
1	3
2	6

x	y
0	0
4	1.2
8	2.4

x	y
2	0.4
4	0.8
6	1.2

x	y
2	1.4
4	1.8
6	2.2

x	y
0	0
1	0.4
2	0.8

x	y
2	−0.8
4	0.4
6	1.6

x	y
0	0
4	28
8	56

x	y
0	4
2	2
4	0

x	y
4	0.4
6	0.6
10	1.0

x	y
5	−1.5
10	−3.0
15	−4.5

x	y
4	−0.4
12	2.8
16	4.4

x	y
2	1.4
3	2.1
4	2.8

x	y
4	4.5
6	5.2
7	5.9

x	y
4	3.2
8	6.4
12	9.6

x	y
2	5
4	10
8	20

x	y
3	8.4
6	16.8
9	25.2

x	y
6	12.2
12	28.4
18	44.6

x	y
6	20.4
10	34
12	40.8

x	y
−1	−12
−2	−24
−3	−36

x	y
−1	−8
0	0
1	8

x	y
−1	4
0	0
1	−4

x	y
−1	4
0	1
1	−2

x	y
1	6
2	12
7	42

x	y
8	44.8
9	50.4
15	84.0

The shaded regions form the letter "V."

Lesson 1.5

1. S **2.** O_1 **3.** L **4.** U **5.** T **6.** I

7. O_2 **8.** N

Lesson 1.6

1. $x + y \leq 15$

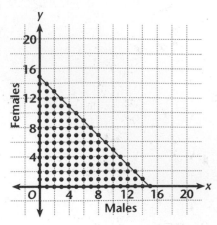

2. $2x + 2y \leq 120$

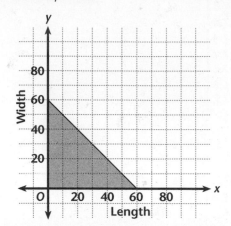

3. $12x + 12y \leq 100$

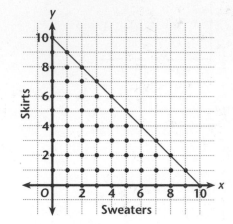

4. $x + y \leq 10$

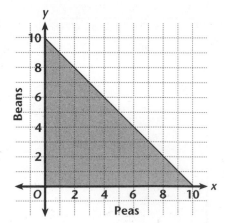

ANSWERS

5. $24x + 30y \le 400$

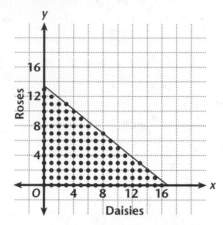

6. $3x + 5y \ge 1500$

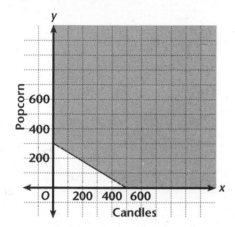

Technology — Chapter 1

Lesson 1.1

1. linear **2.** nonlinear **3.** linear

4. linear **5.** linear **6.** nonlinear

7. nonlinear **8.** nonlinear **9.** nonlinear

10. linear

Lesson 1.2

1. 2.5 **2.** 2.5 **3.** 1 **4.** 2 **5.** −7

6. −2 **7.** 2 **8.** −7 **9.** a

Lesson 1.3

1. 100 differs from the predicted score by more than 15%.

2. 60, 68 (for $h = 2$), 70 (for $h = 2$ and $h = 3$), 74, 80, 94, and 95 differ from the predicted scores by less than 5%.

3. none **4.** 100 seems to be an outlier.

5. average: 94.25; predicted: 91.652. The predicted value is less than the average.

6. average: 69.25; predicted: 74.504. The predicted value is greater than the average.

7.

H	S	5.716*H+57.356	DIFF
1	60	63.072	−3.072
1	55	63.072	−8.072
2	68	68.788	−0.788
2	70	68.788	1.212
2	100	68.788	31.212
3	65	74.504	−9.504
3	70	74.504	−4.504
3	74	74.504	−0.504
3	68	74.504	−6.504
4	70	80.220	−10.220
4	80	80.220	−0.220
4	88	80.220	7.780
5	92	85.936	6.064
5	75	85.936	−10.936
6	85	91.652	−6.652
6	94	91.652	2.348
6	100	91.652	8.348
6	98	91.652	6.348
7	95	97.368	−2.368

8. −0.032

Lesson 1.4

1. yes; 0.5 **2.** no **3.** yes; 1.6 **4.** yes; 1.7

5. no **6.** yes; 0.7 **7.** yes; −1.2 **8.** no

ANSWERS

9. All entries in column A are positive. If the constant of variation is positive, then all entries in column B would be positive. If the constant variation is negative, then all entries in column B would be negative. Since the entries in column B alternate, there is no constant of variation.

Lesson 1.5

1. 1.5   **2.** 10.3   **3.** −1.7   **4.** −5.0

5. no solution   **6.** 6.0   **7.** −5.5   **8.** 9.2

9. no solution   **10.** −16.2

Lesson 1.6

1. $x < -2$ **2.** $x \geq 4$

3. no solution   **4.** all real numbers

5. $x \geq 5$ **6.** $x > 4.5$ **7.** $x \geq 0$ **8.** $x > -6$

9. $0 \leq x \leq 3$ **10.** all real numbers

11. $x \geq 25.72$ **12.** $x \geq 2$

Lesson Activities — Chapter 1

Lesson 1.1

1. x represents the number of months; y represents the amount of growth in inches

2. Answers may vary. Yes. As each consecutive x-value increased by 1, the corresponding y-values increase by 0.5, illustrating an increasing linear relationship.

3. Choose an x-value (number of months), then read the y-value and add your current hair length.

4. It depends on the person's original hair length.

5. Answers will vary but should include the fact that it takes 2 years to grow 12 inches or 1 foot.

6–7. Answers will vary.

Lesson 1.2

1. $d = \frac{5}{4}\ell$ **2.** $d = \frac{1}{8}\ell$ **3.** $d = 3\ell$ **4.** $d = \frac{1}{4}\ell$

5.

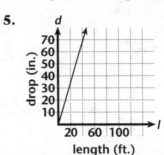

6.

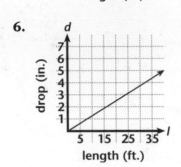

7.

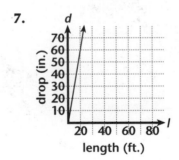

8.

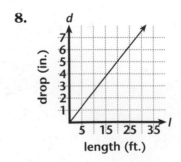

9. $h = \frac{1}{10}\ell$ **10.** Answers will vary.

ANSWERS

Lesson 1.3

1.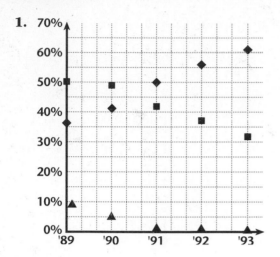

2. positive   **3.** negative   **4.** negative

5. 0.997   **6.** −0.989   **7.** −0.948

8. $y = 6.4x + 36.2$ **9.** $y = -4.5x + 51.2$

10. $y = -2.2x + 7.8$

11. 113%. No, the CD sales cannot exceed 100%.

12. One would expect the total percent of CDs, tapes, and vinyl records for all years to be 100%, but it is about 95%. The remaining percent may be other recording forms such as 8-track tapes.

Lesson 1.4

1. 1 mm ≈ 0.03937 in.   **2.** 39.37 × 55.67

3. 27.83 × 39.37   **4.** 19.69 × 27.83

5. 13.90 × 19.69   **6.** 9.84 × 13.90

7. 6.93 × 9.84   **8.** 4.92 × 6.93

9. 3.46 × 3.92   **10.** 2.44 × 3.46

11. 1.73 × 2.44   **12.** 1.22 × 1.73

13. The new length is equal to the previous poster's width and the new width is half the previous poster's length.

14.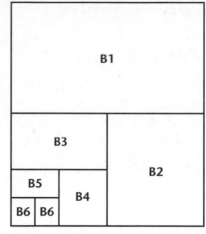

15. Answers will vary.

Lesson 1.5

1. $y = 5x + 75$ **2.** $y = 8x$

3. 25 shirts for $200   **4.** 15 shirts for $150

5. 200 shirts   **6.** $y = 4.5x + 105$

7. 30 shirts   **8.** 180 shirts

9. Answers may vary. Sample answer: Custom Clothes requires only 180 shirts (20 less) be sold for the same $525 profit.

10. 60 shirts

Lesson 1.6

1. $y \geq x - 2$; $y \leq -x + 2$; $y \leq \frac{1}{3}x + 2$; $y \geq -\frac{1}{3}x - 2$

2. The kite is a quadrilateral with pairs of consecutive sides equal in length: Its diagonals are perpendicular bisectors of each other.

3. Answers will vary but should be in the form $a \leq x \leq -6$ where $a \leq -6$.

4. Answers will vary.

ANSWERS

- -

Assessment — Chapter 1

Assessing Prior Knowledge 1.1

1–2.

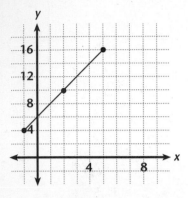

3. 6

Quiz 1.1

1.

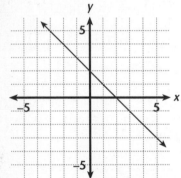

2.

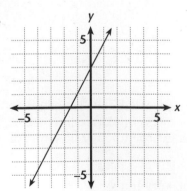

3. $2y = 4x + 6$ is increasing because as the x-values get larger so do the y-values.

4. $c = 15 + 3p$

Assessing Prior Knowledge 1.2

1.

x	−2	−1	0	1	2
y	−3	−1	1	3	5

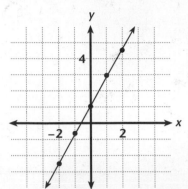

2. Slope is rise over run, or y-change divided by x-change.

3. No.

Quiz 1.2

1. $m = 0, b = 10$ **2.** $m = 2, b = -10$

3. $m = -15, b = 0$

4. slope is undefined, no y-intercept

5. $m = 2, b = -1$ **6.** $m = -1, b = 0$

7. $m = -1$ **8.** $y = -x + 4$ **9.** $y = 5x + 2.5$

10. $y = -x$ **11.** $y = -2$

Assessing Prior Knowledge 1.3

1. The graph is a line with a slope of 2 and a y-intercept of 4; positive

2. A line with positive slope rises from left to right; a line with negative slope falls from left to right.

ANSWERS

Quiz 1.3

1.

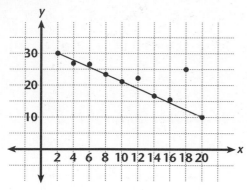

2. negative

3. An outlier is a point that lies noticeably outside the area around the line of best fit. (18, 25) is an outlier.

4. About 7

Mid-Chapter Assessment

1. c **2.** b **3.** c **4.** b

5.

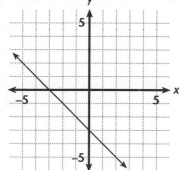

6.

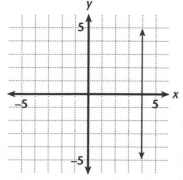

7. $v = 20 + 5n$ **8.** positive

9. (2, 3.5), (6,1) **10.** around 0.6

Assessing Prior Knowledge 1.4

1. 8, 12, 16

2. The y-value is 4 times the x-value. **3.** 1.5

Quiz 1.4

1. $k = 48.2, d = 48.2t$ **2.** $k = 300, d = 300t$

3. $k = \frac{5}{8}, d = \frac{5}{8}t$ **4.** $k = -50, d = -50t$

5. $c = 0.1$ **6.** $d = 24$ **7.** $s = 2p$ **8.** $2.5\,h$

Assessing Prior Knowledge 1.5

1. $x = 2$ **2.** $45x - 30$

Quiz 1.5

1.

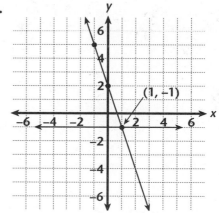

2.

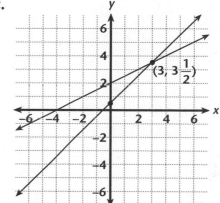

ANSWERS

- -

3. $x = 2$ **4.** $r = \frac{3V}{\pi h}$ **5.** $m = \frac{y - b}{x}$

6. $14 - 2x - 6 = 2x$; $8 - 2x = 2x$;
$8 - 2x + 2x = 2x + 2x$; $8 = 4x$;
$\frac{8}{4} = \frac{4x}{4}$; $x = 2$

Assessing Prior Knowledge 1.6

1. x is less than five.

2.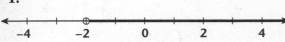

Quiz 1.6

1.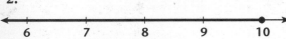

2.

3.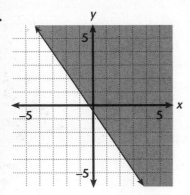

4.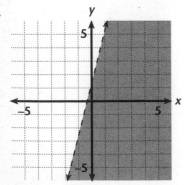

5. $y \le -2x + 2$

Chapter Assessment, Form A

1. a   **2.** b   **3.** c   **4.** a   **5.** d   **6.** b   **7.** a

8. c   **9.** d   **10.** b   **11.** b   **12.** d   **13.** a

14. c   **15.** d   **16.** b

Chapter Assessment, Form B

1. $m = \frac{1}{3}$, $b = -6$

2.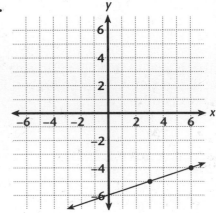

3. Answers will vary. Possible answers are
$y = 5x + 2$ and $y^2 = 4$.

4. $y = -3x - 2$ **5.** $y = x - 5$

6. group of points whose line of best fit has a
slope of -1

7. group of randomly scattered points

8. positive correlation   **9.** no correlation

10.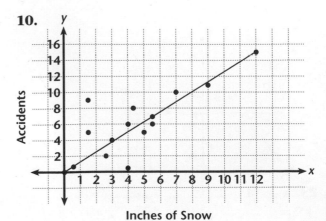

11. $(1.5, 8), (4, 0)$ **12.** positive

ANSWERS

13. $k = -2.8$ 14. $t = 22.5$ 15. $p = 350b$

16. $7000   17. $x = \frac{1}{3}$

18.

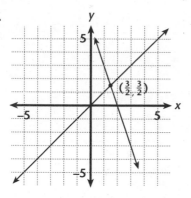

19.

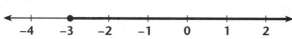

20.

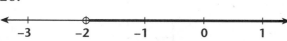

21.

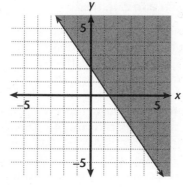

22.

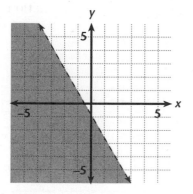

23. $x = -4$

Alternative Assessment — Chapter 1

Form A

1. Answers will vary.

2. Answers will vary.

Numbers of cans	100	200	300
Amount of refund	$3.50	$7.00	$10.50
Numbers of cans	400	500	
Amount of refund	$14.00	$17.50	

3. A constant difference in consecutive x-values results in a constant difference in y-values in the table. The data is linearly related.

4. $y = 0.035x$

5. The class gets $15 for 500 cans. The class needs to recycle 15,000 cans to get $525.

6. The linear equation is a direct variation (containing the point $(0, 0)$) with the constant of variation of 0.035, the amount of refund for 1 can.

Score Point 4: Distinguished

The student demonstrates a comprehensive understanding of linear relationships. The student uses perceptive, creative, and complex mathematical reasoning throughout the task. He or she is able to use sophisticated, precise, and appropriate mathematical language throughout the task. Theoretical knowledge is apparent and applied to concrete situations as the student successfully demonstrates a comprehensive understanding of core concepts.

ANSWERS

Score Point 3: Proficient

The student demonstrates a broad understanding of linear relationships. The student uses perceptive mathematical reasoning throughout the task. He or she is able to use precise and appropriate mathematical language most of the time. Theoretical knowledge is apparent and applied to concrete situations as the student attempts to draw conclusions based on his or her investigations.

Score Point 2: Apprentice

The student demonstrates an understanding of the use of linear relationships. The student uses mathematical reasoning at times during the task. He or she uses appropriate mathematical language some of the time. Student attempts to apply theoretical knowledge to the task, but may not always be able to draw conclusions from his or her investigation.

Score Point 1: Novice

The student demonstrates a basic understanding of the use of linear relationships. The student uses mathematical reasoning. He or she uses appropriate mathematical language some of the time. Theoretical knowledge is extremely weak and many responses are irrelevant or illogical. He or she may fail to follow directions and has great difficulty in communicating his or her responses.

Score Point 0: Unsatisfactory

Student fails to make an attempt to complete the task and his or her responses are just an attempt to fill the page or restate the problem.

Form B

1. Answers will vary. Possible answer:
   x is the number of adventure software games
   y is the number of sports games
   $10x + 15y \leq 120$

2. $y \leq -\frac{2}{3}x + 8$ or $x \leq -\frac{3}{2}y + 12$

3. Only whole numbers are possible.

4.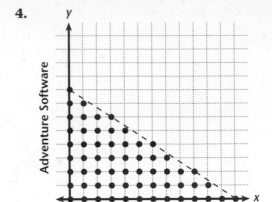

5. The cost of 6 adventure games and 4 sports games is $120.

6.

Number of sports games	0	3	6	9	12
Number of adventure software	8	6	4	2	0

Score Point 4: Distinguished

The student demonstrates a comprehensive understanding of solving linear inequalities. The student uses perceptive, creative, and complex mathematical reasoning throughout the task. He or she is able to use sophisticated, precise, and appropriate mathematical language throughout the task. Theoretical knowledge is apparent and applied to concrete situations as the student successfully demonstrates a comprehensive understanding of core concepts.

Score Point 3: Proficient

The student demonstrates a broad understanding of solving linear inequalities. The student uses perceptive mathematical reasoning most of the time. He or she is able to use precise and appropriate mathematical language most of the time. Theoretical knowledge is apparent and applied to concrete situations as the student attempts to draw conclusions based on his or her investigations.

ANSWERS

Score Point 2: Apprentice

The student demonstrates an understanding of solving linear inequalities. The student uses mathematical reasoning at times during the task. He or she uses appropriate mathematical language some of the time. Student attempts to apply theoretical knowledge to the task, but may not always be able to draw conclusions from his or her investigation.

Score Point 1: Novice

The student demonstrates a basic understanding of solving linear inequalities. The student uses mathematical reasoning. He or she uses appropriate mathematical language some of the time. Theoretical knowledge is extremely weak and many responses are irrelevant or illogical. He or she may fail to follow directions and has great difficulty in communicating his or her responses.

Score Point 0: Unsatisfactory

Student fails to make an attempt to complete the task and his or her responses are just an attempt to fill the page or restate the problem.

Practice & Apply — Chapter 2

Lesson 2.1

1. $\frac{19}{28}$ **2.** $\frac{3}{7}$ **3.** $-\frac{3x}{4}$ **4.** $2a$ **5.** $\frac{2}{3}$

6. $18n + \frac{5}{15}$ **7.** $\frac{4}{3(x+3)}$ **8.** $\frac{y-6}{6}$ **9.** $\frac{y}{5}$

10. $\frac{23a-8}{20}$ **11.** $\frac{4-21c}{18}$ **12.** $\frac{4c}{27}$

13. Commutative property of addition

14. Associative property of multiplication

15. Distributive property

16. Multiplicative inverse

17. $\frac{y(y+2)}{4}$ square meters

Lesson 2.2

1. x^7 **2.** d^6 **3.** y^2 **4.** $20x^3y^4$ **5.** $-4a^2b$

6. $3^{\frac{5}{2}}$ **7.** $16n^{12}$ **8.** $\frac{12}{xy^3}$ **9.** $\frac{4}{9y^6}$ **10.** $7^{5.2}$

11. $\frac{c^6}{27}$ **12.** 3^4 **13.** $V = y^3$

14. $y = \frac{x^5}{x^2}$

x	2	1	0	−1	−2
y	8	1	und	−1	−8

$y = x^3$

x	2	1	0	−1	−2
y	8	1	0	−1	−8

If $x \neq 0$, the graph and tabular values are the same because $\frac{x^5}{x^2}$ is equal to x^3. If $x = 0$, then $\frac{x^5}{x^2}$ is undefined.

Lesson 2.3

1. yes **2.** no **3.** yes **4.** yes **5.** no

6. yes **7.** no **8.** function

9. not a function **10.** function

11. domain: all real numbers; range: all real numbers; function

12. domain: all real numbers; range: all real numbers such that $y \geq 3$; function

13. domain: all real numbers such that $x \geq 0$ range: all real numbers; not a function

14. domain: all real numbers except 0; range: all real numbers except 0; function

15. $y = 0.75x$ **16.** Check students' work.

17. \$93.75

Lesson 2.4

1. 27 **2.** −1 **3.** 1 **4.** $5a - 3$ **5.** −9

6. $-\frac{1}{5}$ **7.** $5x + 5h - 3$ **8.** $x^2 - 2x - 9$

9. $\frac{1}{x-2}$ **10.** $15r - 13$ **11.** $4n^2 - 20n + 15$

12. 5 **13.** $f(d) = \pi d$ **14.** $3\pi x$

15. Check students' work.

16. D: all real numbers   **17.** $(1, 4)$ **18.** $R: y \le 4$

19. $C(x) = \$0.20x + 25$ **20.** $\$100$

Lesson 2.5

1. $f(x) = 5x + 7$ **2.** $f(x) = \frac{1}{3}x - 1$

3. $f(x) = -4x + 2$ **4.** $f(x) = x + 6$

5. $2; f(x) = 2x + 7$ **6.** $-\frac{3}{4}; f(x) = -\frac{3}{4}x + \frac{11}{4}$

7. $\frac{2}{3}; f(x) = \frac{2}{3}x - 2$ **8.** $-1; f(x) = -x + 7$

9. $f(x) = -2x - 1$ **10.** $f(x) = \frac{3}{4}x + \frac{11}{2}$

11. $f(x) = 5x - 14$ **12.** $f(x) = -\frac{1}{2x} + 5$

13. decreasing   **14.** increasing

15. increasing   **16.** decreasing

17. $f(x) = 2x - 2$ **18.** $f(x) = -3$

19. $f(x) = -x$ **20.** $(2, 0)$ **21.** $\left(\frac{5}{2}, 0\right)$

22. $(-3, 0)$ **23.** none

Lesson 2.6

1. $x^2 - 3x + 1; -x^2 - 3x + 1$

2. $x^2 - 3x + 5; x^2 + 3x - 9$

3. $\frac{1 + 4x^2}{x}; \frac{1 - 4x^2}{x}$ **4.** $6\sqrt{x}; -4\sqrt{x}$

5. $\frac{2x^2 - 4x}{x - 3}; \frac{-2x^2 + 8x}{x - 3}$ **6.** $\frac{2x + 6}{x + 2}; \frac{4x + 6}{x + 2}$

7. $12x^2; 3$ **8.** $4x^2 - 15x + 11; \frac{4x - 11}{x - 1}$

9. $\frac{2x - 4}{x}; \frac{1}{8x^2 - 16x}$ **10.** $x^2 - 7x; \frac{x}{x - 7}$

11. $-3^{\frac{1}{2}}x^{\frac{3}{2}}; -\frac{\sqrt{3}}{\sqrt{x}}$ or $-\frac{\sqrt{3x}}{x}$

12. $(x + 4)^{\frac{3}{2}}(x - 4)^{\frac{1}{2}}; -\frac{\sqrt{x - 4}}{\sqrt{x + 4}}$ or $-\frac{\sqrt{x^2 - 16}}{x + 4}$

13. domain: all real numbers; range: all real numbers such that $y \ge 4$

14. domain: all real numbers; range: all real numbers such that $y \ge 2$

15. domain: all real numbers; range: all real numbers

16. domain: all real numbers except $\frac{1}{2}$; range: all real numbers such that $y \le -1.6$ or $y \ge 2.6$

17. (0.667); coordinates of f and $-f$ are the y-opposites of each other.

18. $f(x) = 475 + 0.07x; \$615$

Enrichment — Chapter 2

Lesson 2.1

1. $\frac{1}{4} + \frac{1}{8}$ **2.** $\frac{1}{2} + \frac{1}{4} + \frac{1}{10} + \frac{1}{20}$ **3.** $\frac{1}{5} + \frac{1}{15}$

4. $\frac{1}{2} + \frac{1}{3} + \frac{1}{12}$ **5.** $\frac{1}{9} + \frac{1}{27}$

6. $\frac{1}{2} + \frac{1}{3} + \frac{1}{18} + \frac{1}{30}$ **7.** $\frac{1}{2} + \frac{1}{40} + \frac{1}{80}$

8. $\frac{1}{2} + \frac{1}{9}$ **9.** $\frac{1}{2} + \frac{1}{4} + \frac{1}{13} + \frac{1}{10} + \frac{1}{26} + \frac{1}{65}$

10. $\frac{1}{3} + \frac{1}{75}$ **11.** $\frac{1}{2} + \frac{1}{13} + \frac{1}{7} + \frac{1}{82}$

12. $\frac{1}{2} + \frac{1}{5} + \frac{1}{20} + \frac{1}{40}$ **13.** $\frac{1}{3} + \frac{1}{12}$

14. $\frac{1}{2} + \frac{1}{4} + \frac{1}{10}$ **15.** $\frac{1}{5} + \frac{1}{35}$

16. $\frac{1}{3} + \frac{1}{10} + \frac{1}{30}$ **17.** $\frac{1}{5} + \frac{1}{10} + \frac{1}{50}$

18. $\frac{1}{2} + \frac{1}{10} + \frac{1}{13} + \frac{1}{26} + \frac{1}{65}$

Lesson 2.2

1. $\$128.01$ **2.** $\$253.35$ **3.** $\$634.12$

4. $\$634.87$ **5.** $\$1610.32$ **6.** $\$1972.34$

7. $\$204.35$ **8.** $\$4373.33$ **9.** $\$5093.81$

ANSWERS

10. $10,437.12 **11.** $269.16 **12.** $270.48

Lesson 2.3

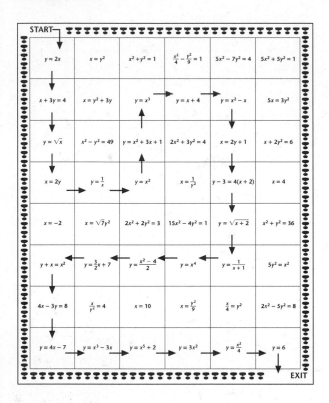

Lesson 2.6

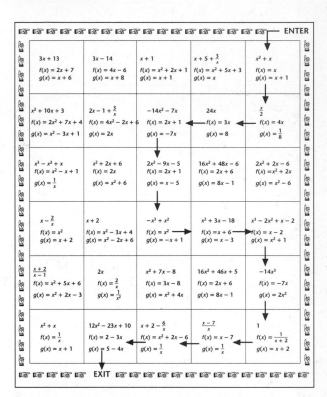

Technology — Chapter 2

Lesson 2.1

1. 1.41421356 **2.** 2.23606798

3. 4.24264069 **4.** 9.79795897

5. 10.04987562 **6.** 17.32050808

7. 7.07106781 **8.** -3.46410162

9. 0.317837245 **10.** 8.36308110

Lesson 2.2

1.

1	A	B	C	D	E
2	P	R	N	T	A
3	1250	0.075	1	0	1250.00
4				1	1343.75
5				2	1444.53
6				3	1552.87

Lesson 2.4

1. 3 **2.** $\frac{3}{4}$ **3.** -2 **4.** $2x$ **5.** $-2x$

6. $2x$ **7.** $2x$ **8.** $4x + 3$ **9.** $-6x + 4$

Lesson 2.5

1. S **2.** L **3.** O **4.** P **5.** I **6.** N

7. G **8.** L **9.** I **10.** N **11.** E **12.** S

13. C **14.** A **15.** N **16.** S **17.** L

18. A **19.** N **20.** T

21. Sloping lines can slant.

ANSWERS

2.

1	A	B	C	D	E
2	P	R	N	T	A
3	1250	0.075	2	0	1250.00
4				1	1345.51
5				2	1448.31
6				3	1558.97
7				4	1678.09
8				5	1806.30

3.

1	A	B	C	D	E
2	P	R	N	T	A
3	1250	0.075	3	0	1250.00
4				1	1346.11
5				2	1449.62
6				3	1561.08
7				4	1681.11
8				5	1810.37
9				6	1949.57
10				7	2099.48
11				8	2260.91

4.

1	A	B	C	D	E
2	P	R	N	T	A
3	1250	0.075	4	0	1250.00
4				1	1346.42
5				2	1450.28
6				3	1562.15
7				4	1682.64
8				5	1812.44
9				6	1952.24
10		.		7	2102.83
11				8	2265.03
12				9	2439.74
13				10	2627.94
14				11	2830.64

5.

1	A	B	C	D	E
2	P	R	N	T	A
3	1250	0.055	1	0	1250.00
4				1	1318.75
5				2	1391.28
6				3	1467.80

6.

1	A	B	C	D	E
2	P	R	N	T	A
3	1250	0.065	1	0	1250.00
4				1	1331.25
5				2	1417.78
6				3	1509.94

7.

1	A	B	C	D	E
2	P	R	N	T	A
3	1250	0.075	1	0	1250.00
4				1	1343.75
5				2	1444.53
6				3	1552.87

8.

1	A	B	C	D	E
2	P	R	N	T	A
3	1250	0.085	1	0	1250.00
4				1	1356.25
5				2	1471.53
6				3	1596.61

9. The more frequently interest is computed in a year, the greater the amount in the account at a given time.

10. The greater the interest rate, the more will be in the account at a given time.

ANSWERS

Lesson 2.3

1.

	A	**B**	**C**	**D**
1	STEP 1	STEP 2	STEP 3	STEP 4
2	−2	4	10	9
3	−1	1	4	3
4	0	0	0	−1
5	1	1	−2	−3
6	2	4	−2	−3

2.

	A	**B**	**C**	**D**
1	STEP 1	STEP 2	STEP 3	STEP 4
2	−2	12	20	16
3	−1	3	7	3
4	0	0	0	−4
5	1	3	−1	−5
6	2	12	4	0

3.

	A	**B**	**C**	**D**
1	STEP 1	STEP 2	STEP 3	STEP 4
2	−2	8	2	0
3	−1	2	−1	−3
4	0	0	0	−2
5	1	2	5	3
6	2	8	14	12

4.

	A	**B**	**C**	**D**
1	STEP 1	STEP 2	STEP 3	STEP 4
2	−2	4	2	1
3	−1	1	0	−1
4	0	0	0	−1
5	1	1	2	1
6	2	4	6	5

5.

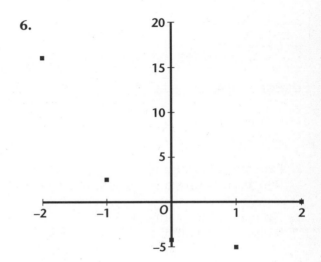

6.

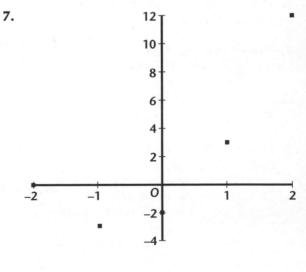

7.

ANSWERS

8.

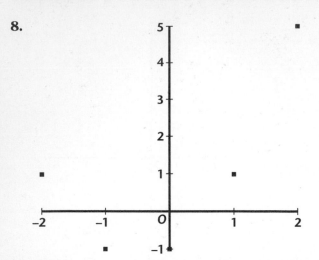

9. −3.25

Lesson 2.4

1. Domain: all real numbers
   Range: all real numbers

2. Domain: all real numbers
   Range: all nonnegative real numbers

3. Domain: all real numbers
   Range: all real numbers

4. Domain: all real numbers
   Range: all nonnegative real numbers

5. Domain: all real numbers
   Range: all real numbers

6. Domain: all real numbers
   Range: all nonnegative real numbers

7. Domain: all real numbers
   Range: all real numbers

8. Domain: all real numbers
   Range: all nonnegative real numbers

9. Domain: all real numbers except −1, 0, and 1
   Range: all real numbers except 0

10. Domain: all real numbers except −2, −1, 0, 1, and 2
    Range: all real numbers except 0

11. If $f(x) = x^n$, the domain is all real numbers. If n is a positive odd integer, the range is all real numbers. The graph is a sideways S. If n is a positive even integer, the range is all nonnegative real numbers. The graph is U-shaped and opens upward.

12. Domain: all real numbers except those real numbers which make the denominator 0. Range: all real numbers except 0. The graph consists of U shapes that alternately open up and down. At the left and right ends of the graph are branches that are above the x-axis and open upward.

Lesson 2.5

1.

	A	B	C
1	X	F(X)	SLOPE
2	0	0.0	
3	1	0.25	0.25
4	2	1.00	0.75
5	3	2.25	1.25
6	4	4.00	1.75

2.

	A	B	C
1	X	F(X)	SLOPE
2	0	0.00	
3	1	0.2	0.2
4	2	0.8	0.6
5	3	1.8	1.0
6	4	3.2	1.4

3.

	A	B	C
1	X	F(X)	SLOPE
2	0	0.00	
3	1	0.15	0.15
4	2	0.60	0.45
5	3	1.35	0.75
6	4	2.40	1.05

ANSWERS

4.

	A	B	C
1	X	F(X)	SLOPE
2	0	0.00	
3	1	0.1	0.1
4	2	0.4	0.3
5	3	0.9	0.5
6	4	1.6	0.7

5.

	A	B	C
1	X	F(X)	SLOPE
2	0	0.00	
3	1	−1.0	−1
4	2	−4.0	−3
5	3	−9.0	−5
6	4	−16.0	−7

6.

	A	B	C
1	X	F(X)	SLOPE
2	0	0.00	
3	1	−0.5	−0.5
4	2	−2.0	−1.5
5	3	−4.5	−2.5
6	4	−8.0	−3.5

7.

	A	B	C
1	X	F(X)	SLOPE
2	0	0.00	
3	1	−0.25	−0.25
4	2	−1.00	−0.75
5	3	−2.25	−1.25
6	4	−4.00	−1.75

8.

	A	B	C
1	X	F(X)	SLOPE
2	0	0.00	
3	1	−0.1	−0.1
4	2	−0.4	−0.3
5	3	−0.9	−0.5
6	4	−1.6	−0.7

9. If $f(x) = ax^2$ and a is positive, the graph curves up to the right. The slopes of the line segments are positive and increasing.

10. If $f(x) = ax^2$ and a is negative, the graph curves down to the right. The slopes of the line segments are negative and decreasing.

Lesson 2.6

1. Domain: all real numbers
Range: all real numbers less than or equal to 0.125.

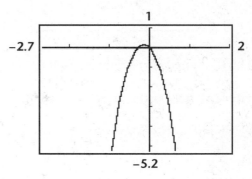

2. Domain: all real numbers
Range: all real numbers greater than or equal to −1

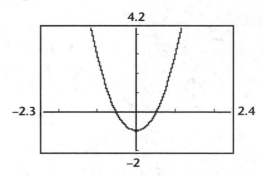

ANSWERS

3. Domain: all real numbers
   Range: all real numbers greater than or equal to 0

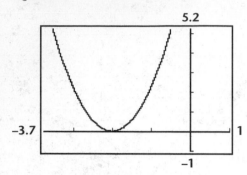

4. Domain: all real numbers
   Range: all real numbers greater than or equal to -6.900833

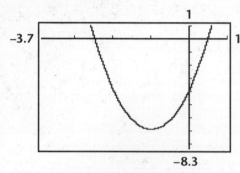

5. The U-shaped curve opens upward.

6. The U-shaped curve opens downward.

7. 0   8. -1 9. 9   10. -4

11. The y-intercept is bd.

Lesson Activities — Chapter 2

Lesson 2.1

1–4. Answers will vary.

5. Check students' responses. Possible answer:

MAY

S	M	T	W	T	F	S
		1	2	3	4	5	6
7	8	9	10	11	12	13
14	15	16	17	18	19	20
21	22	23	24	25	26	27
28	29	30	31			

6. Add the dates along either diagonal of the square, or add the two opposite corner dates and double the results.

Lesson 2.2

1. 4 and 6; odd: 4, even: 6

2. 6

x^1	x^2	x^3	x^4	x^5	x^6	Units Digits
1	1	1	1	1	1	1
2	4	8	16	32	64	2,4,8,6
3	9	27	81	243	729	3,9,7,1
4	16	64	256	1024	4096	4,6
5	25	125	625	3125	15625	5
6	36	216	1296	7776	46656	6
7	49	343	2401	16807	117649	7,9,3,1
8	64	512	4096	32768	262144	8,4,2,6
9	81	729	6561	59049	531441	9,1
10	100	1000	10000	100000	100000	0

3. 4 digits is the longest cycle.   4. 1   5. 50

6. $5^6 = 15625$, $6^5 = 7776$. Therefore, 5^6 is larger.

Lesson 2.3

1. 25 ft   2. 45 ft

3. Check students' graphs.   4. $d = 5t^2$

5. The equation is a function because there is a 1-1 mapping between time and distance.

6. Domain: $t \geq 0$; Range: $d \geq 0$ 7. 18,000 ft

Lesson 2.4

1. 2424 calories   2. 3600 calories

3. Student answers should be in the form $C(t) = 60K$ where K = number of calories used per minute for a particular activity given in the table.

4. Check students' graphs.

5. Domain: $t \geq 0$ 6. Range: calories ≥ 0

7. The relationship is linear.   8. 576 calories

ANSWERS

Lesson 2.5

1. Student tables will vary.

2. Amount of flow = rate of flow × time

3. Student answers will be dependent on the rate of flow.

4. The constant rate of change is the rate of flow.

5. Yes. The amount of flow and time passed are related by an equation of the form $y = kx$.

6. The slope of the function is the flow rate.

7. 0   8. increasing

Lesson 2.6

1. $Y_1 = -0.85x + 11.97$; $Y_2 = -0.66x + 8.84$; $Y_3 = -14x + 224$

2. 4   3. fewer

4. The dependent variable is the number of discs purchased.

5. Yes. Each function has a negative slope.

6. $Y_4 = -15.51x + 244.81$

7. The domain for each function is the price of the CDs. The range for each function is the number of discs purchased. The range of Y_4 is the union of the ranges of Y_1, Y_2, and Y_3.

Assessment — Chapter 2

Assessing Prior Knowledge 2.1

1. 3.2   2. $-2\frac{2}{3}$ 3. $-\frac{1}{15}$ 4. -1

Quiz 2.1

1. 9   2. $\frac{23}{18}$ 3. $\frac{1}{6}$ 4. $\frac{3}{16}$ 5. 2   6. $\frac{1}{x}$

7. $\frac{1}{6}$ 8. x 9. $\frac{3x^2 + 9x}{x^2}$ 10. $\frac{-x - 8}{6}$

11. $\frac{4}{y^2 - y}$ 12. $\frac{5}{6x}$

Assessing Prior Knowledge 2.2

1. 27   2. 49   3. 32   4. 64

Quiz 2.2

1. 16   2. $\frac{1}{4}$ 3. 27   4. $\frac{1}{2}$ 5. y^2

6. $\frac{-125y^9}{64x^6}$ 7. $\frac{-3z^3}{4xy^4}$ 8. $\frac{3}{b}$ 9. $15x^3y$

10. $\frac{-1}{4mn^{10}}$ 11. 27   12. $\frac{72x^6}{y^2}$

Assessing Prior Knowledge 2.3

1.

x	-3	-2	-1	0	1	2	3
y	19	9	3	1	3	9	19

2. Yes, -3 and 3 correspond to 19; -2 and 2 correspond to 9; -2 and 2 correspond to 9; and -1 and 1 correspond to 3.

Quiz 2.3

1. not a function; more than one y value for $x = 1$

2. function   3. function

4. function   5. function

6. not a function; more than one y value for $x = 2$ and $x = -1$

7. function

Mid-Chapter Assessment

1. b   2. a   3. d   4. c   5. $\frac{37}{35}$ 6. $\frac{-1}{3}$

7. $\frac{7x - 5}{10}$ 8. $\frac{2}{9}$ 9. $\frac{x}{6}$ 10. 16

11. $\frac{4z^8}{9x^7y^6}$ 12. $-2a^3bc^2$

Assessing Prior Knowledge 2.4

1. $y = 3$ 2. $y = -3$

ANSWERS

Quiz 2.4

1. 0   **2.** $a^2 + 4a$ **3.** -6

4. $a^2 + b^2 - 2ab - 6a + 6b + 3$

5.

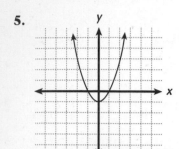

6. domain = all real numbers

7. range = $f(x) \geq -1$ **8.** $(0, -1)$

9. independent: x; dependent: $f(x)$

Assessing Prior Knowledge 2.5

1. -2 **2.** 4   **3.** -0.5 **4.** 0.625

Quiz 2.5

1. $f(x) = \frac{1}{2}x + 2$; increasing

2. $f(x) = -x - 1$; decreasing

3. $f(x) = 2x$; increasing

4. $f(x) = -x + 7$; decreasing

5. $f(x) = -\frac{2}{3}x + 2$; decreasing

6. $f(x) = 3$; neither

Assessing Prior Knowledge 2.6

1. $2x + 3$ **2.** $x - 2$ **3.** $x^2 - x - 6$

4. $2x^2 + 9x - 5$

Quiz 2.6

1. D = all real numbers; R = all real numbers

2. D = all real numbers; $R = y \leq 1$

3. D = all real numbers; $R = y \leq \frac{9}{4}$

4. D all real numbers except $x \neq \frac{1}{3}$;
R = all real numbers

5. D = all real numbers except $x \neq 1$, $x \neq -1$;
R = all real numbers

6. D = all real numbers; R = all real numbers

7. $(f \div g)(x) = \frac{x - 1}{2x^2}$; $(g \cdot f)(x) = 8x^4(x - 1)$

8. $(f \div g)(x) = 2x^2(x - 3)^2$; $(g \cdot f)(x) = 2x^2$

9. $(f - g)(x) = 2(2x^3 - x + 4)$;
$\left(g \cdot \frac{1}{f}\right)(x) = \frac{(x - 4)}{2x^3}$

10. $(f - g)(x) = 3(-x^2 + 7x + 1)$;
$\left(g \cdot \frac{1}{f}\right)(x) = \frac{x^2 - 4x - 2}{3x - 1}$

Chapter Assessment, Form A

1. d   **2.** b   **3.** b   **4.** a   **5.** b   **6.** c   **7.** d

8. a   **9.** d   **10.** d   **11.** d   **12.** c   **13.** d

14. b

Chapter Assessment, Form B

1. $f(x) = 2x - 1$ **2.** $f(x) = -\frac{1}{2}x + 4$

3. function   **4.** not a function

5. not a function   **6.** $f(x) = -2x - 1$

7. $f(x) = \frac{1}{3}x + 2$ **8.** $\frac{x - 1}{3}$ **9.** $x + 4$

10. $\frac{x^2 - x + 2}{x^2 - 1}$ **11.** $\frac{6x^2z^3}{5y^2}$

12. function   **13.** not a function

ANSWERS

14.

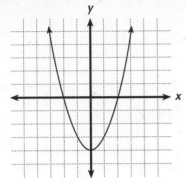

15. D = all real numbers   **16.** $R = y \geq -4$

17. $b^2 + 2b - 3$ **18.** $(0, -4)$

19. not a function   **20.** function

21. $\dfrac{1}{3x(x^3 - 27)}$

22. D = all real numbers except $x \neq 0$ and $x \neq 3$

23.

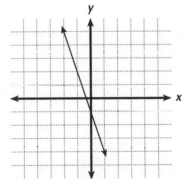

24.

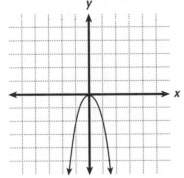

Alternative Assessment — Chapter 2

Form A

1. Answers will vary. Any real number may represent a. Any integers may represent m and n.

2. The equation $a^m a^n = a^{mn}$ is false. The true equation is $a^m a^n = a^{m+n}$. $\dfrac{a^n}{a^m} = a^{n-m}$ is true.

3. $\dfrac{1}{3xy}$; Answers may vary. $a^{-1} = \dfrac{1}{a}$ of $(a^m)^n = a^{mn}$ and $a^{-1} = \dfrac{1}{a}$.

4. 1; Answers may vary. $a^0 = 1$ or $(a^m)^n = a^{mn}$ and $a^0 = 1$.

5. Answers will vary. $(1 + 2)^{-1} \neq -1 + \dfrac{1}{2}$.

6. Answers will vary. $1 + 2^{-1} \neq \dfrac{1}{3}$.

7. Answers will vary. $1^{-1} + 2^{-1} = 1 + \dfrac{1}{2}$.

8. Answers will vary. $(1 + 2)^{-1} = \dfrac{1}{3}$.

9. Equations 7 and 8 are true.

10. For all real numbers a, $(1 + a)^{-1} = \dfrac{1}{1 + a}$.

Score Point 4: Distinguished

The student demonstrates a comprehensive understanding of properties of exponents. The student uses perceptive, creative, and complex mathematical reasoning throughout the task. He or she is able to use sophisticated, precise, and appropriate mathematical language throughout the task. Theoretical knowledge is apparent and applied to concrete situations as the student successfully demonstrates a comprehensive understanding of core concepts throughout the task.

ANSWERS

Score Point 3: Proficient

The student demonstrates a broad understanding of properties of exponents. The student uses perceptive mathematical reasoning most of the time. He or she is able to use precise and appropriate mathematical language most of the time. Theoretical knowledge is apparent and applied to concrete situations as the student attempts to draw conclusions based on his or her investigations.

Score Point 2: Apprentice

The student demonstrates an understanding of properties of exponents. He or she uses mathematical reasoning at times during the task. He or she uses appropriate mathematical language some of the time. Student attempts to apply theoretical knowledge to the task but may be able to draw conclusions from his or her investigation.

Score Point 1: Novice

The student demonstrates a basic understanding of properties of exponents. He or she uses appropriate mathematical language some of the time. Theoretical knowledge is extremely weak and many responses are irrelevant or illogical. He or she may fail to follow directions and has great difficulty in communicating his or her responses.

Score Point 0: Unsatisfactory

Student fails to make an attempt to complete the task and his or her responses are just an attempt to fill the page or restate the problem.

Form B

1. The domain of the sum function consists of those values of x common to the domains of f and g. The range of the sum function is the sum of the range values of f and g for each x. $(f + g)(x) = x^2 + 4x + 4$.

2. $(f - g)(x) = -x - 2x$. $(f - g)(x) = f(x) - g(x)$ because you can use the graphs of two functions to find their difference by subtracting at each value of the independent variable, the values of the two dependent variables.

3. The domain of the product function consists of those values of x common to the domains of f and g. The range of the product function is found by multiplying, at each value of the independent variable, the values of the two dependent variables. $(f \cdot g)(x) = x^3 + 5x^2 + 8x + 4$. The product function is commutative.

4. The domain of the quotient function consists of those values of x common to the domains of f and g. Any value that makes the denominator, $g(x)$, 0 is excluded from the quotient function. $\dfrac{f(x)}{g(x)} = \dfrac{x + 2}{(x + 1)(x + 2)}$ $= \dfrac{1}{x + 1}$ where $x \neq -1$ or -2.

Score Point 4: Distinguished

The student demonstrates a comprehensive understanding of operations with functions. The student uses perceptive, creative, and complex mathematical reasoning throughout the task. He or she is able to use sophisticated, precise, and appropriate mathematical language throughout the task. Theoretical knowledge is apparent and applied to concrete situations as the student successfully demonstrates a comprehensive understanding of core concepts throughout the task.

Score Point 3: Proficient

The student demonstrates a broad understanding of operations with functions. The student uses perceptive mathematical reasoning most of the time. He or she is able to use precise and appropriate mathematical language most of the time. Theoretical knowledge is apparent and applied to concrete situations as the student attempts to draw conclusions based on his or her investigations.

Score Point 2: Apprentice

The student demonstrates an understanding of operations with functions. He or she uses mathematical reasoning at times during the task. He or she uses appropriate mathematical language some of the time. Student attempts to apply theoretical knowledge to the task but may be able to draw conclusions from his or her investigation.

ANSWERS

Score Point 1: Novice

The student demonstrates a basic understanding of operations with functions. He or she uses appropriate mathematical language some of the time. Theoretical knowledge is extremely weak and many responses are irrelevant or illogical. He or she may fail to follow directions and has great difficulty in communicating his or her responses.

Score Point 0: Unsatisfactory

Student fails to make an attempt to complete the task and his or her responses are just an attempt to fill the page or restate the problem.

Practice & Apply — Chapter 3

Lesson 3.1

1. symmetric with respect to x-axis

2. symmetric with respect to y-axis

3. symmetric with respect to y-axis

4. $A'(-1, -2)$; $A'(1, 2)$

5. $B'(0, 2)$; $B'(0, -2)$ 6. $C'(-2, 0)$; $C'(2, 0)$

7. $A'(-2, 1)$, $B'(-3, 2)$, $C'(-1, 3)$

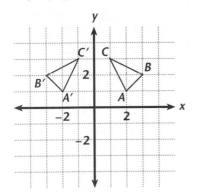

8. The image of triangle $A'B'C'$ is a reflection of the pre-image triangle ABC over the y-axis. See graph for Exercise 7.

9. $(-5, 4)$ 10. $(-2, 0)$ 11. $(3, 8)$

12. Two points are symmetric with respect to the line $y = x$ if the x- and y-coordinates are interchanged.

13. $A'(0, 3)$, $B'(0, 5)$, $C'(2, 5)$, $D'(2, 3)$

14. The image square $A'B'C'D'$ is a reflection of the pre-image square $ABCD$ over the line $y = x$.

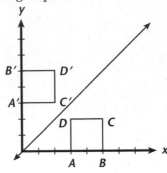

Lesson 3.2

1. $\{(3, 4), (2, 2), (1, 0), (0, -2)\}$; function

2. $\{(-5, -3), (1, -2), (3, 1), (6, 2)\}$; function

3. $\{(-2, 6), (-1, 5), (3, 0), (-2, 5)\}$; not a function

4. $\{(8, 6), (6, 4), (3, 2), (5, 1)\}$; function

5. yes   6. yes   7. no   8. no

9. $f^{-1}(x) = \frac{x - 4}{7}$ 10. $f^{-1}(x) = 5x + 30$

11. $f^{-1}(x) = \frac{2x + 1}{3}$ 12. $3; 4$

13. $f^{-1}(x) = \frac{x - 4}{3}$ 14. $\frac{1}{3}; -\frac{4}{3}$

15. The slopes of f and f^{-1} are inverses of each other.

Lesson 3.3

1. The domain and range of f and g are the set of real numbers.

2. $f \circ g = 3x + 1$; $g \circ f = 3x - 1$ 3. no

4. The domain and range of $f \circ g$ and $g \circ f$ are the set of real numbers.

5. $2x + 2$ 6. $2x + \frac{5}{2}$ 7. $4x + 9$ 8. $x - 1$

9. -4 10. 1 11. $\frac{5}{2}$ 12. 13

ANSWERS

13. $2\pi r$; The result is another form for the circumference of a circle.

14. $9x^2 - 6x - 1$ 15. $-3x^2 + 7$ 16. $4x^2 - 2$

17. $-6x + 2$ 18. $36x^2 - 12x - 1$ 19. 14

20. 2   21. 4

22. Answer may vary.
Sample answer: $g(x) = x + 5, f(x) = x^2$

23. Answers may vary.
Sample answer: $g(x) = 3x, f(x) = x - 4$

24. $f(2) + 3 = 7$

25. $g(2) = 8, f(8) = 8 - 1 = 7$

26. Definition of composition.

Lesson 3.4

1. -0.5 2. 0.75   3. 8   4. 2   5. 3.34

6. 2.6   7. $x = 8$ or $x = 1$

8. $x = 3$ or $x = -7$ 9. $x = 0$ or $x = -4$

10. $x = -2$ or $x = 6$ 11. $x = 4$ or $x = -4$

12. $x = 2$ or $x = -4$ 13. $\varnothing$

14. $x = -6$ or $x = 9$ 15. $x = 8$ or $x = -16$

16. $x = \frac{8}{3}$ 17. $x = 0$ or $x = -4$

18. $x = -\frac{1}{2}$ or $x = 3$

19. domain: all real numbers; range: all real numbers ≥ 2

20. domain: all real numbers; range: all real numbers ≥ -1

21. domain: all real numbers; range: all real numbers ≤ 0

22. Check students' work.

23. It is the same graph translated 3 units right and 3 units down.

24. $f(x) = |x - 1|$ 25. $f(x) = -|x + 2|$

Lesson 3.5

1. 4   2. -2 3. 6   4. -4 5. -5

6. -4 7. 1   8. -7 9. 3   10. -3

11. 13   12. -3 13. 2   14. 0   15. -2

16. 7   17. 6   18. -6 19. -14 20. -2

21. 2   22. -10 23. -11 24. 14

25. -3 26. 2   27. -3 28. 0

29. Check students' work.

30. $-2[x]$ steps 2 units vertically at intervals of 1 unit. $[-2x]$ steps 1 unit vertically at intervals of 1/2 unit.

31. 2   3   4   5   6   7

32.

33. For each x-value there is exactly one y-value.

34. rounding-up function

35. $C(t) = [2t - 1] + 2$ 36. $17

37. The cost would increase without bound.

38. $2.72

Lesson 3.6

1. $y = x + 9$ 2. $y = -\frac{2}{3}x + \frac{2}{3}$

ANSWERS

3. $y = -x + 1$ 4. $y = 3x + 17$

5. $y = -3x + 2$ 6. $y = -x + 2$

7. Answers may vary.
   Sample answer: $x = 2 - 3t; y = 3 + t$

8. $x(t) = 125t$
   $y(t) = 30t$

9. $x = 37{,}500$ ft; $y = 9000$ ft   10. $y = \frac{6}{25}x$

11. ≈ 35 ft   12. ≈ 1.4 s

13. ≈ 70 ft   14. ≈ 2.8 s

Enrichment — Chapter 3

Lesson 3.1

1. I   2. M   3. A   4. G   5. E

Lesson 3.2

The letter f

Lesson 3.3

1. T   2. H   3. E   4. T   5. W   6. O

7. A   8. R   9. E   10. I   11. N   12. V

13. E   14. R   15. S   16. E   17. S

Lesson 3.4

Sample answers are given.

1. impossible

2. $x = 1, y = 2; x = -1, y = -2$

3. $x = 1, y = -2; x = 6, y = -3$

4. $x = 4, y = 2; x = -1, y = 3$

5. $x = 4, y = 1; x = -1, y = -2$

6. $x = 2, y = -3; x = 1, y = 4$

7. $x = 4, y = 1; x = 1, y = 4$

8. impossible

9. $x = 6, y = 2; x = 6, y = 4$

10. $x = 8, y = 2; x = 7, y = 3$

11. $x = 0, y = 1; x = 1, y = 0$

12. $x = -1, y = 6; x = -2, y = 3$

13. $x = -3, y = 5; x = 0, y = 2$

14. $x = 7, y = 3; x = 9, y = 2$

15. $x = 4, y = 0; x = 8, y = 5$

16. impossible

17. $x = 2, y = 1; x = 3, y = -1$

18. impossible

19. $x = 1, y - 2; x = 3, y = 4$

20. $x = 3, y = 7; x = -2, y = -3$

21. impossible   22. impossible

23. $x = 1, y = 2; x = 4, y = -2$

24. $x = 6, y = 3; x = -8, y = -2$

ANSWERS

Lesson 3.5

1.

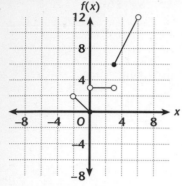

2.

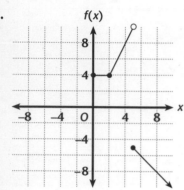

3.

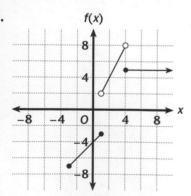

4.

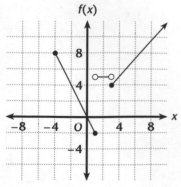

5.

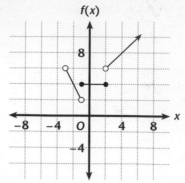

6.

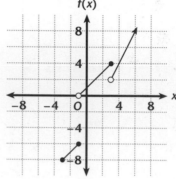

Lesson 3.6

1. $y = x + 4$ **2.** $y = \frac{1}{2}x + \frac{9}{2}$

3. $y = -\frac{2}{3}x + \frac{13}{3}$ **4.** $y = -6x + 32$

5. $y = \frac{9}{8}x + \frac{17}{8}$ **6.** $y = -\frac{4}{3}x + \frac{17}{3}$

7. $y = \frac{2}{3}x - \frac{44}{3}$ **8.** $y = \frac{1}{6}x + \frac{5}{3}$

9. $y = -\frac{9}{8}x - \frac{1}{8}$

ANSWERS

Technology — Chapter 3

Lesson 3.1

1. $x = 0$

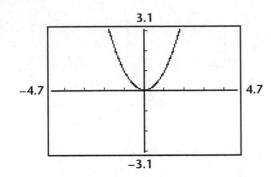

2. $x = 0$

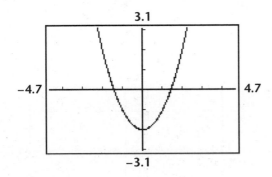

3. $x = -1$

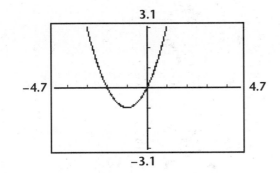

4. $x = 2$

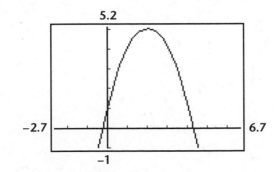

5. $x = 3$

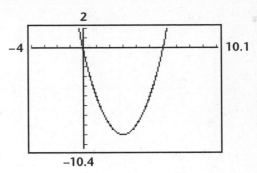

6. $x = -1$

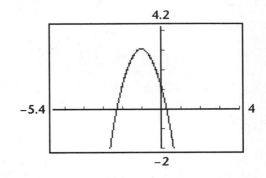

7. $(0.3333, -0.0741)$

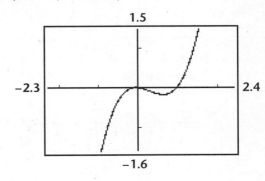

8. $(0, 3)$

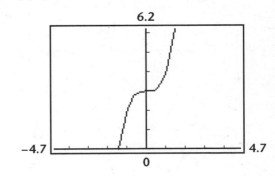

9. $(0.5, 0)$

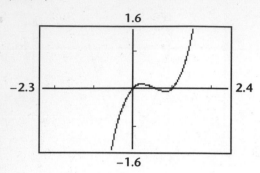

10. The graph would contain $(r, -s)$ as well as (r, s). Thus, the graph would not pass the vertical-line test. So, it cannot be the graph of a function.

11. Answers will vary. A sample is given. Let $a = 1$, $b = -8$, and $c = 0$.

Lesson 3.2

1. True, since the reflection of $f(x) = x$ in the line $y = x$ is itself.

2. The statement is false, since any graph of such a function does not pass the horizontal line test. The graphs of $f(x) = x^2$, $g(x) = 2x^2$, and $h(x) = 3x^2$ are shown along with horizontal lines that cross the graphs in more than one point.

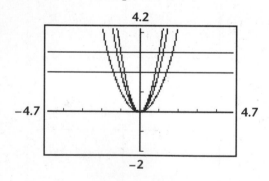

3. The statement is false, since any graph of such a function does not pass the horizontal line test. The graphs of $f(x) = -x^2$, $g(x) = -2x^2$, and $h(x) = -3x^2$ are shown along with horizontal lines that cross the graphs in more than one point.

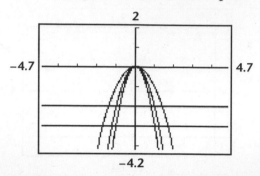

4. The statement is true as long as a is not equal to 0. The graphs of $f(x) = x^3$, $g(x) = -2x^3$, and $h(x) = 3x^3$ are shown along with horizontal lines that cross the graphs in no more than one point.

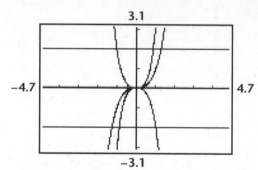

5. True. The graphs of $f(x) = \frac{1}{x}$, $g(x) = \frac{2}{x}$, and $h(x) = \frac{3}{x}$ are shown along with the graph of $f(x) = x$.

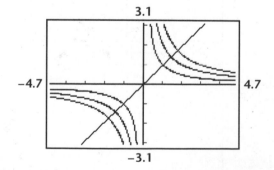

ANSWERS

6. The statement is true as long as *a* is not equal to 0. The graphs of $f(x) = x^2$, $g(x) = -2x^2$, and $h(x) = 3x^2$ are shown along with horizontal lines that cross the graphs in no more than one point.

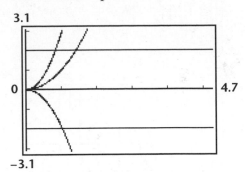

7. The statement is true as long as *a* is not equal to 0. The graphs of $f(x) = x^2$, $g(x) = -2x^2$, and $h(x) = 3x^2$ are shown along with horizontal lines that cross the graphs in no more than one point.

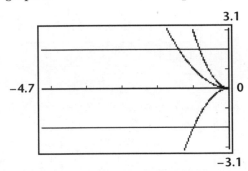

8. False. Let $m = 1$ and $b = 0$. The graphs of *f* and *g* are shown. They are not reflections of one another in the line $y = x$.

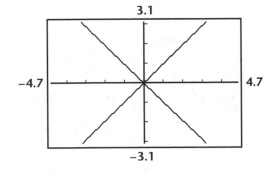

Lesson 3.3

1. Yes, the graphs coincide.

2. Yes, the graphs coincide.

3. The graphs from Exercises 1 and 2 suggest that composition of simple power functions is commutative.

4. The graphs from Exercises 1 and 2 suggest that composition of simple power functions is another power function.

5. If *m* and *n* are both even, the graph is a U-shaped curve opening upward. The greater *mn* is, the flatter the U is at the bottom.

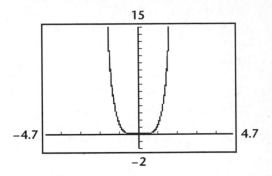

6. If *m* and *n* are both odd, the graph is a sideways S-shaped curve passing through the origin. The greater *mn* is, the flatter the S is at the origin.

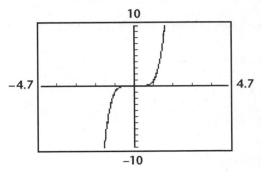

7.

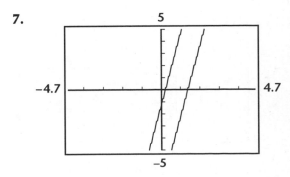

ANSWERS

8.

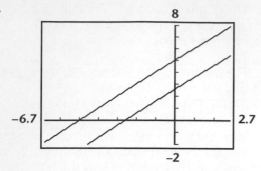

9. Since the graphs from Exercises 7 and 8 are straight lines, you can conclude that the composition of two linear functions is linear.

10. Since the graphs from Exercises 7 and 9 are different straight lines, you can conclude that the composition of two linear functions is not commutative.

Lesson 3.4

1. $x = -4$ or 2   **2.** $x = 0$ **3.** $x = 0.333333$

4. no solution   **5.** no solution

6. $x \approx -0.9789474$ or 1.1052632

7. all real numbers   **8.** $x \geq 0$

9. $x < -0.5263158$ or $x > 0.73684211$

10. $0 < x < 1$

Lesson 3.5

1.

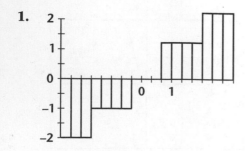

2.

3.

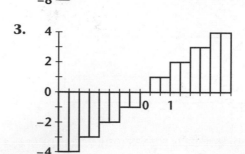

4.

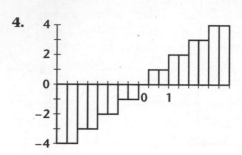

5.

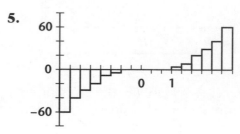

6.

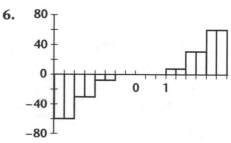

7.

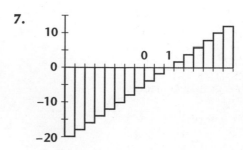

ANSWERS

8.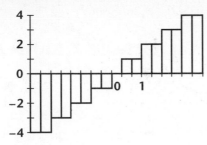

9.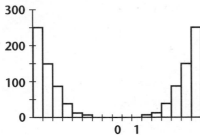

10.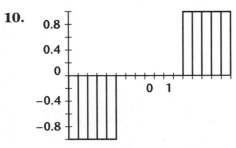

11. The composition of *f* with itself *n* times is *f*.

Lesson 3.6

1.

1	A	B	C
2	T	X	Y
3	0.0	−4	−4
4	0.5	0	0
5	1.0	4	4

2.

1	A	B	C
2	T	X	Y
3	0.00000000	0.00000000	0.00000000
4	0.14285714	0.57142857	0.57142857
5	0.28571429	1.14285714	1.14285714
6	0.42857143	1.71428571	1.71428571
7	0.57142857	2.28571429	2.28571429
8	0.71428571	2.85714286	2.85714286
9	0.85714286	3.42857143	3.42857143
10	1.00000000	4.00000000	4.00000000

3.

1	A	B	C
2	T	X	Y
3	0.0	−4	−4
4	0.5	−2	0
5	1.0	0	4

4.

1	A	B	C
2	T	X	Y
3	0.0	0	0
4	0.2	2	0
5	0.4	4	0
6	0.6	6	0
7	0.8	8	0
8	1.0	10	0

5.

1	A	B	C
2	T	X	Y
3	0.00000000	0	0
4	0.16666667	0	1
5	0.33333333	0	2
6	0.50000000	0	3
7	0.66666667	0	4
8	0.83333333	0	5
9	1.00000000	0	6

ANSWERS

6.

1	A	B	C
2	T	X	Y
3	0.00000000	−4.0000000	−2
4	0.16666667	−1.6666667	−1
5	0.33333333	0.66666667	−2.22E-16
	0.50000000	3.00000000	1
	0.66666667	5.33333333	2
	0.83333333	7.66666667	3
6	1.00000000	10.00000000	4

7.

1	A	B	C
2	T	X	Y
3	0.00	−6	6
4	0.25	−3	3
5	0.50	0	0
6	0.75	3	−3
7	1.00	6	−6

8.

1	A	B	C
2	T	X	Y
3	0.00000000	1.50000000	0.00000000
4	0.16666667	1.25000000	0.83333333
5	0.33333333	1.00000000	1.66666667
6	0.50000000	0.75000000	2.50000000
7	0.66666667	0.50000000	3.33333333
8	0.83333333	0.25000000	4.16666667
9	1.00000000	1.6653E-16	5.00000000

9.

1	A	B	C
2	T	X	Y
3	0.00000000	0.0	5.00000000
4	0.16666667	2.5	4.16666667
5	0.33333333	5.0	3.33333333
6	0.50000000	7.5	2.50000000
7	0.66666667	10.0	1.66666667
8	0.83333333	12.5	0.83333333
9	1.00000000	15.0	5.5511E-16

10. $R(2.66666667, 3.26666667)$

Lesson Activities — Chapter 3

Lesson 3.1

1. yes **2.** yes **3.** yes

4. Possible answer: $y = -x$

5. Check students' designs.

6. Answers will vary.

Lesson 3.2

1. Check students' graphs.

2.

	$f(x)$		$f^{-1}(x)$	
t	x	y	x	y
−10	−10	−15	−15	−10
−5	−5	−5	−5	−5
0	0	5	5	0
5	15	15	15	5
10	10	25	25	10

3. −15; −10 **4.** (−5, −5) **5.** b; a

6. For each element in the domain there is one element in the range.

7. $f^{-1}(x) = 0.5x - 2.5$ or $\frac{x}{2} - \frac{5}{2}$.

8. No; $g(x)$ fails the horizontal line test.

9. Possible answer: use Tmin = 0

Lesson 3.3

1. $f(x) = -\frac{5}{3}x + 95$ **2.** $g(x) = 6x$

3. $f(g(x)) = -10x + 95$

4. The domain is Hours Working. The range is Test Score.

146 **Answers**   **HRW Advanced Algebra**

5. Check students' graphs.   **6.** 80

7. No. Possible answer: Negotiate working 1.5 or fewer hours on one day and making up the time on another day.

Lesson 3.4

1. Final Function Rule: $f(x) = -|x - 4| - 6$

2. Final Function Rule: $f(x) = \left|\dfrac{x}{2} + 2\right| - 3$

Lesson 3.5

1.

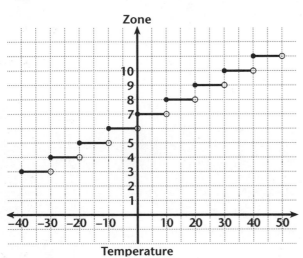

2. The greatest integer function.

3. $y = [0.1x + 6.5]$

4. Answers will vary.   **5.** Answers will vary.

Lesson 3.6

1. The parametric representations are graphed at the same time.

2. The coordinates of points in the top and bottom lines are reversed. They are reflections of one another in the center line, $y = x$, and inverse functions.

3. $f(x) = 2x - 8$

4. $f^{-1}(x) = 0.5x + 4$ or $f^{-1}(x) = \dfrac{x}{2} + 4$

5. Domain: all reals; Range: all reals

6. Domain: all reals; Range: all reals

7. The inverse is a function; it passes the vertical line test.

8. The inverse is not a function.

9. Possible answer: use domain where $x \geq -2$

Assessment — Chapter 3

Assessing Prior Knowledge 3.1

1–4.

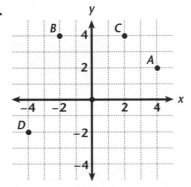

5 A and B **6.** C and D

Quiz 3.1

1. $A'(3, -4); B'(1, 2); C'(-5, 0)$

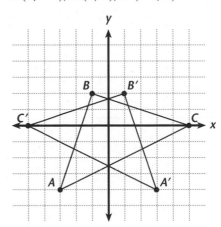

2. $A'(-3, 4); B'(-1, -2); C'(5, 0)$

3. $y = x; (-7, 9); (2, -10); (0, 0); (8, -10);$ x-axis

ANSWERS

Assessing Prior Knowledge 3.2

1.

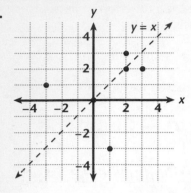

Quiz 3.2

1. $\{(-8, -2), (-1, -1), (7, 3), (27, 3)\}$

2. No inverse

3.

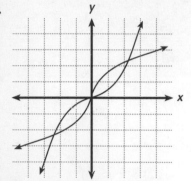

4.

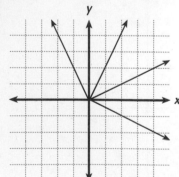

5. $f(x) = \frac{1}{2}(x - 1)$ **6.** $f(x) = \frac{3}{2}x - \frac{9}{2}$

7. $f(x) = \frac{1}{x} + 1$ or $\frac{x + 1}{x}$, $x \neq 0$

Assessing Prior Knowledge 3.3

1. $g(-1) = 3$, $g(2) = 0$, $g(0) = 0$

2. $h(-1) = 0$, $h(2) = 1.5$, $h(0) = 0.5$

Quiz 3.3

1. $2x^2 + 6$ **2.** $4x^2 + 16x + 17$ **3.** 14

4. 5   **5.** 0   **6.** 17

7. $f \circ g = -2x - 1; g \circ f = -2x + \frac{1}{2}$

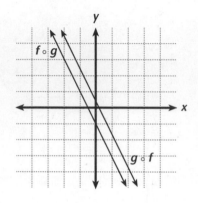

8. $-2x - \frac{1}{2}$

Mid-Chapter Assessment

1. c   **2.** c   **3.** a   **4.** d   **5.** b

6. $y = -\frac{2}{3}x - \frac{8}{3}$

7. $A'(1, -6), B'(4, 4), C'(-4, 2), D'(-3, 2)$

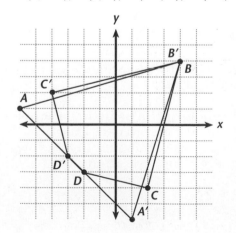

ANSWERS

Assessing Prior Knowledge 3.4

1–2.

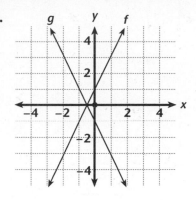

10.

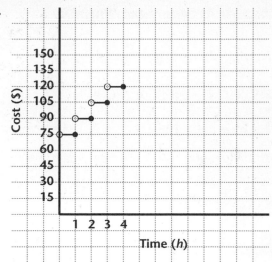

Quiz 3.4

1. −1 **2.** 4 **3.** −6 **4.** $f(x) = |x| - 3$

5. $f(x) = -2|x|$ **6.** $f(x) = -\frac{1}{2}|x + 2|$

7. $x = 4\frac{1}{3}$ or $x = -3$ **8.** $x = -6$ or $x = 4$

9. $x = \frac{7}{2}$ of $x = \frac{-7}{2}$

11. $120

Assessing Prior Knowledge 3.6

1. $y = -2x - 11$ **2.** $y = \frac{1}{2}x + 4$ **3.** $y = 2$

Assessing Prior Knowledge 3.5

1.

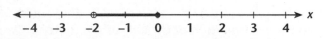

2.

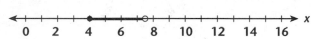

Quiz 3.6

1.

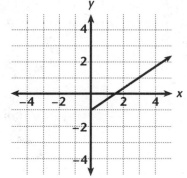

Quiz 3.5

1. −2 **2.** 1 **3.** −1 **4.** 11 **5.** −1

6. 1 **7.** −4 **8.** 5 **9.** $C(t) = 75 + 15\lceil t \rceil$

2. $f(x) = \frac{1}{6}x + 1$ **3.** $f(x) = 4x - 2$

4. $f(x) = 15x$

5. radius: $x(t) = 15 - 1t$
circumference: $y(t) = 94.2 - 6.28t$

6. 62.8 mm

Chapter Assessment, Form A

1. d **2.** c **3.** b **4.** d **5.** a **6.** d **7.** a

8. a **9.** d **10.** c **11.** b **12.** d

ANSWERS

Chapter Assessment, Form B

1. $\{(\frac{2}{3}, -2), (-\frac{1}{3}, 1), (-1, 3)\}$

2. $A'(-2, -6); B'(2, 6); C'(-4, 1)$

3. $\{(0.5, 0), (-14.5, 3), (-4.5, 1), (5.5, -1)\}$

4. Range: $g(x) \geq + 1$ 5. Range: $h(x) \leq 0$

6. Range: $f(x) \geq -3$ 7. $x = \frac{13}{7}$ and $x = 1$

8. -4.3 9. 16 10. -3 11. -1.9

12. $x^2 - 2$ 13. -1 14. $2x + 1$ 15. 1

16. $\sqrt{2x - 3}$ 17. 1

18.

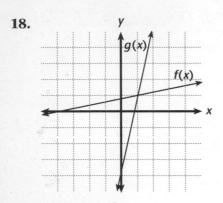

19. $g \circ f = 5\left(\dfrac{x + 4}{5}\right) - 4$

$\qquad = x + 4 - 4$

$\qquad = x$

$\qquad f \circ g = \dfrac{(5x - 4) + 4}{5}$

$\qquad = \dfrac{5x}{5}$

$\qquad = x$

20. $f(x) = 10x - 10$

21. volume: $x(t) = 100t$
    diameter: $y(t) = 1.8t$

22. 14.4 cm

Alternative Assessment — Chapter 3

Form A

1.

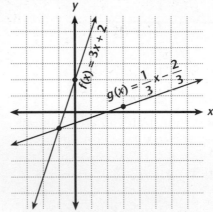

2. The domain of f is the range of g. The range of f is the domain of g.

3. f and g are inverse functions. For all points (a, b) on the graph of f (b, a) is a point on the graph of g.

4. The equation of the axis of symmetry is $I(x) = x$.

5. $f \circ g = x$ and $g \circ f = x$. $f \circ g$ and $g \circ f$ are equal to the equation of the axis of symmetry.

6. The composition of any function and its inverse is x.

Score Point 4: Distinguished

The student demonstrates a comprehensive understanding of inverse functions and the composition of functions. The student uses perceptive, creative, and complex mathematical reasoning throughout the task. He or she is able to use sophisticated, precise, and appropriate mathematical language throughout the task. Theoretical knowledge is apparent and applied to concrete situations as the student successfully demonstrates a comprehensive understanding of core concepts throughout the task.

ANSWERS

Score Point 3: Proficient

The student demonstrates a broad understanding of inverse functions and the composition of functions. The student uses perceptive mathematical reasoning most of the time. He or she is able to use precise and appropriate mathematical language most of the time. Theoretical knowledge is apparent and applied to concrete situations as the student attempts to draw conclusions based on his or her investigations.

Score Point 2: Apprentice

The student demonstrates an understanding of inverse functions and the composition of functions. He or she uses mathematical reasoning at times during the task. He or she uses appropriate mathematical language some of the time. Student attempts to apply theoretical knowledge to the task but may be able to draw conclusions from his or her investigation.

Score Point 1: Novice

The student demonstrates a basic understanding of inverse functions and the composition of functions. He or she uses appropriate mathematical language some of the time. Theoretical knowledge is extremely weak and many responses are irrelevant or illogical. He or she may fail to follow directions and has great difficulty in communicating his or her responses.

Score Point 0: Unsatisfactory

Student fails to make an attempt to complete the task and his or her responses are just an attempt to fill the page or restate the problem.

Form B

1.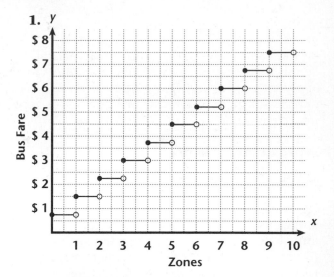

2. Answers may vary. The graph looks like a series of steps.

3. The function is a rounding up function.

4. $f(x) = 0.75\lceil x \rceil$

5. The function represents a scalar transformation of the rounding-up function.

Score Point 4: Distinguished

The student demonstrates a comprehensive understanding of applications of step functions. The student uses perceptive, creative, and complex mathematical reasoning throughout the task. He or she is able to use sophisticated, precise, and appropriate mathematical language throughout the task. Theoretical knowledge is apparent and applied to concrete situations as the student successfully demonstrates a comprehensive understanding of core concepts throughout the task.

Score Point 3: Proficient

The student demonstrates a broad understanding of applications of step functions. The student uses perceptive mathematical reasoning most of the time. He or she is able to use precise and appropriate mathematical language most of the time. Theoretical knowledge is apparent and applied to concrete situations as the student attempts to draw conclusions based on his or her investigations.

ANSWERS

Score Point 2: Apprentice

The student demonstrates an understanding of applications of step functions. He or she uses mathematical reasoning at times during the task. He or she uses appropriate mathematical language some of the time. Student attempts to apply theoretical knowledge to the task but may be able to draw conclusions from his or her investigation.

Score Point 1: Novice

The student demonstrates a basic understanding of applications of step functions. He or she uses appropriate mathematical language some of the time. Theoretical knowledge is extremely weak and many responses are irrelevant or illogical. He or she may fail to follow directions and has great difficulty in communicating his or her responses.

Score Point 0: Unsatisfactory

Student fails to make an attempt to complete the task and his or her responses are just an attempt to fill the page or restate the problem.

Cumulative Assessment Free-Response Grids

Exercise _____

Exercise _____

Exercise _____

Exercise _____

Exercise _____

Exercise _____

Exercise _____

Exercise _____

Exercise _____

PORTFOLIO HOLISTIC SCORING GUIDE

An individual portfolio is likely to be characterized by some, but not all, of the descriptors for a particular level. Therefore, the overall score should be the level at which the appropriate descriptors for a portfolio are clustered.

		NOVICE ○	APPRENTICE ○	PROFICIENT ○	DISTINGUISHED ○
PROBLEM SOLVING	Understanding/Strategies	• Indicates a basic understanding of problems and uses strategies	• Indicates an understanding of problems and selects appropriate strategies	• Indicates a broad understanding of problems with alternate strategies	• Indicates a comprehensive understanding of problems with efficient, sophisticated strategies
	Execution/Extensions	• Implements strategies with minor mathematical errors in the solution without observations or extensions	• Accurately implements strategies with solutions, with limited observations or extension	• Accurately and efficiently implements and analyzes strategies with correct solutions, with extension	• Accurately and efficiently implements and evaluates sophisticated strategies with correct solutions and includes analysis, justifications, and extensions
REASONING		• Uses mathematical reasoning	• Uses appropriate mathematical reasoning	• Uses perceptive mathematical reasoning	• Uses perceptive, creative, and complex mathematical reasoning
MATHEMATICAL COMMUNICATION	Language	• Uses appropriate mathematical language some of the time	• Uses appropriate mathematical language	• Uses precise and appropriate mathematical language most of the time	• Uses sophisticated, precise, and appropriate mathematical language throughout
	Representations	• Uses few mathematical representations	• Uses a variety of mathematical representations accurately and appropriately	• Uses a wide variety of mathematical representations accurately and appropriately; uses multiple representations within some entries	• Uses a wide variety of mathematical representations accurately and appropriately; uses multiple representations within entries and states their connections
UNDERSTANDING/CONNECTING CORE CONCEPTS		• Indicates a basic understanding of core concepts	• Indicates an understanding of core concepts with limited connections	• Indicates a broad understanding of some core concepts with connections	• Indicates a comprehensive understanding of core concepts with connections throughout
TYPES AND TOOLS		• Includes few types; uses few tools	• Includes a variety of types; uses tools appropriately	• Includes a wide variety of types; uses a wide variety of tools appropriately	• Includes all types; uses a wide variety of tools appropriately and insightfully

WORKSPACE/ANNOTATIONS

PERFORMANCE DESCRIPTORS

PROBLEM SOLVING
- Understanding the features of a problem (understands the question, restates the problem in own words)
- Explores (draws a diagram, constructs a model and/or chart, records data, looks for patterns)
- Selects an appropriate strategy (guesses and checks, makes an exhaustive list, solves a simpler but similar problem, works backward, estimates a solution)
- Solves (implements a strategy with an accurate solution)
- Reviews, revises, and extends (verifies, explores, analyzes, evaluates strategies/solutions; formulates a rule)

REASONING
- Observes data, records and recognizes patterns, makes mathematical conjectures (inductive reason)
- Validates mathematical conjectures through logical arguments or counter-examples; constructs valid arguments (deductive reasoning)

MATHEMATICAL COMMUNICATION
- Provides quality explanations and expresses concepts, ideas, and reflections clearly
- Uses appropriate mathematical notation and terminology
- Provides various mathematical representations (modes, graphs, charts, diagrams, words, pictures, numerals, symbols, equations)

UNDERSTANDING/CONNECTING CORE CONCEPTS
- Demonstrates an understanding core concepts
- Recognizes, makes, or applies the connections among the mathematical core concepts to other disciplines, and to the real world

Place an X on each continuum to indicate the degree of understanding demonstrated for each core concept.

	DEGREE OF UNDERSTANDING OF CORE CONCEPTS	
	Basic	Comprehensive with connections
NUMBER		
MATHEMATICAL PROCEDURES		
SPACE & DIMENSIONALITY		
MEASUREMENT		
CHANGE		
MATHEMATICAL STRUCTURE		
DATA: STATISTICS AND PROBABILITY		

PORTFOLIO CONTENTS
- Table of Contents
- Letter to Reviewer
- 5–7 Best Entries

BREADTH OF ENTRIES

TYPES
- INVESTIGATIONS/DISCOVERY
- APPLICATIONS
- NON-ROUTINE PROBLEMS
- PROJECTS
- INTERDISCIPLINARY
- WRITING

TOOLS
- CALCULATORS
- COMPUTER AND OTHER TECHNOLOGY
- MODELS-MANIPULATIVE
- MEASUREMENT INSTRUMENTS
- OTHERS

GROUP ENTRY

The Kentucky Mathematics Portfolio was developed by the Kentucky Department of Education for use by school districts throughout that state.